PENGUIN BOOKS

UP ON THE RIVER

John Madson is well known in the fields of ecology and the outdoors. His articles appear regularly in such national publications as *National Geographic* and *Audubon*. His last book, *Where the Sky Began*, was cited as one of the "most notable books" of 1982 by *The New York Times*.

Up on
the River

JOHN MADSON

Illustrations by Dycie Madson

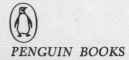

PENGUIN BOOKS

PENGUIN BOOKS
Viking Penguin Inc., 40 West 23rd Street,
New York, New York 10010, U.S.A.
Penguin Books Ltd, Harmondsworth,
Middlesex, England
Penguin Books Australia Ltd, Ringwood,
Victoria, Australia
Penguin Books Canada Limited, 2801 John Street,
Markham, Ontario, Canada L3R 1B4
Penguin Books (N.Z.) Ltd, 182–190 Wairau Road,
Auckland 10, New Zealand

First published in the United States of America by
Nick Lyons Books, Inc., a division of Schocken Books 1985
Published in Penguin Books 1986

LIBRARY OF CONGRESS CATALOGING IN PUBLICATION DATA
Madson, John.
    Up on the river.
    1. Natural history—Mississippi River.   2. Stream
ecology—Mississippi River.   3. Mississippi River.
I. Title.
QH104.5.M5M33   1986   574.5′26323′0977   86-5028
ISBN 0 14 00.8746 X

Printed in the United States of America by
R. R. Donnelley & Sons Company, Harrisonburg, Virginia
Set in Caledonia

Contents

Up on the River

Prologue

Between the Saints

From the mouth of Titus Hollow where we had put in I poled the freight canoe for a mile across the backwaters of the Batchtown flats, along channels twisting through beds of smartweed. We picked our way down the marsh trails in the open sun and morning wind, looking for passage to the River and finally breaking out into the main channel just in time to catch the wheel wash of a northbound towboat.

"My word," said the Englishman. "Oh, my bloody word."

His attention was fixed ahead, on the River, although the stuff behind us was notable enough. Back there the loess-capped headlands of Calhoun County marked the west coast of Illinois—a backlit mosaic of rock, lofty goat prairies, and oak forests reflected in the waters under the bluffs. I could see no buildings or highlines, nor any sign of people: only those headlands with their stone fronts still in shadow, frowning out over the sunlit marsh.

Before us, the great River sliding seaward. I pushed into the current and let it take us, drifting down toward the gray cape of rock that the old Frenchmen had called "Cap au Gris." I waited for Willy Newlands to say something else but he stared at the great brown Mississippi in silence, really being on it for the first time and probably getting that visceral turn that coming suddenly out into the River always seems to trigger.

"Well, here it is in all its turbid glory," I opened. "How about it?"

"A bit bigger than I'd thought," he answered, still not taking his eyes from the River.

"Bigger than most, smaller than some. You've known the Thames and the Rhine; this can't be all that much more."

Willy unfastened himself from the scenery and looked at me thoughtfully. "But it is, you know. It's the Mississippi." He paused, turning back to the River again. "And I'm here."

Before that, I had been studying German under Herr Professor Schmidt. It was scientific German, which is the very worst kind, with its compound word-horrors that only German scientists could devise, or would ever want to.

Herr Schmidt was the archetypal teacher of this awful *wissen-*

schaftlich Schriftdeutsch. He was seldom in the classroom as we straggled in, but would enter precisely on the hour—a short, compact man of middle years, walking briskly and somewhat pigeon-toed and as perfectly perpendicular as it's possible for a man to be without hanging. Invariably clad in a dark suit with sincere necktie and crisp white shirt, he drilled us in technical word roots and pluperfect subjunctives with no more literary emphasis than the log-log decitrig slide rules we used in those ancient days.

One day, however, our translations happened to include a scrap of Thomas Mann. To our surprise, this moved Herr Schmidt to a brief comment on literature—and the revelation that when he had arrived in this country his first act, even before reporting to his new teaching job, was to make a pilgrimage to Hannibal, Missouri, to pay homage to Tom Sawyer and the fabled Mississippi. Within that starched Teutonic bosom beat the heart of a born river rat.

Later, I was a flunky for a German film crew on location in western Illinois. It was a cosmopolitan bunch. The star of our little series was a Norwegian skiing champion known in some circles as "The Golden Stud." The script girl was a stunning Scot out of South Africa, of all places. The head cameraman was Dido Weigert of Munich, ably assisted by his countryman Atzie Hamel.

Dido was an unflappable pro who'd already shared an Emmy for a documentary on the Kremlin. No nonsense with Dido. He wasn't even ruffled that blazing August afternoon when the superbly endowed script girl begged our kind indulgence to strip to her waist. Gentlemen all, we permitted her to work the rest of the day in the comfort of a black lace half-bra.

Anyway, about the only thing that flapped Dido was the nearby Mississippi and our river yarns. Like Willy Newlands and Herr Professor Schmidt, the first American writing he'd ever read was Mark Twain and he'd never really gotten over it. Toward the end of our filming we had a little trouble keeping him on the track and off the River, and he departed vowing to return and make the definitive Mississippi River film for European television. Sooner or later, he'll be back.

And then came Father Patrick Purcell from the Glen of Atherlow in County Tipperary. He was a seasoned outdoorsman. It was said that he'd been something of a salmon poacher back in his friskier days—and when he spoke of his home rivers it was in an affectionate but somewhat guarded manner, with lowered voice. Or maybe it only seemed so. But he did have the feel for rugged places and wild water that ex-poachers never lose, and with his first long look at the mile-wide Mississippi and its alluvial forests he said lovingly: "Ah, 'tis the River itself . . ." He said it as only an Irishman can, which is about as good as it can get.

I have found it interesting and a bit puzzling, looking at the Mississippi in the company of Scots, Englishmen, Irishmen, Canadians, Aussies, Mexicans, Germans, French, and Lebanese. It strikes them in different ways, of course, although I have a hunch that that visceral turn is usually there. For some, there may even be a flash of special insight.

It has been said that to understand Americans one must understand baseball. The trouble is, the nuances of baseball are utterly beyond the grasp of most foreigners. It might be simpler for them to understand something of the Mississippi River, which would certainly constitute a more perceptive encounter with the confused and confusing heart of America than would a visit to Yankee Stadium or Disneyland.

As with any other part of real America, of course, they'll have to look for the Mississippi. Strange, how such a deep and wide and almighty long river can be so hard to find—even for Americans. Especially for Americans. Foreigners may find it more easily than we can, and even after its old storybook spell is broken for them, as it surely will be, they may still have the conviction that this is the quintessential American river, our heartland's fullest gathering of waters, folkways, and manners. Ask any foreigner to name just one American river and see what he says.

If that visitor spends enough time on and around the Mississippi he could end up understanding Americans better than most of us know ourselves. Prey of greed and spoils politics, like us, flowing out of a hopeful past into an aimless and uncertain future, like us, the Great River embodies all our follies, fancies, and glories. Like us—rich, powerful, colorful, polluted, wasted, beloved, feared, serene, brutal, ugly, and beautiful. It is not the American Ganges;

the soul of America is not to be found there. It is only reflected there. And like America as a whole, it is a joyous place once you shrug off its sour detractors and find it for yourself.

You will hear it called "The Great Sewer," the intestinal tract of America's midsection, fit only for commercial traffic and waste disposal.

There is something to that, but the larger truth is that great stretches of the Mississippi are lovely corridors of wildness that still honor original landscapes in what otherwise is a bland monotony of corn, soybeans, and cotton. It is a pity that we have profaned and strictured parts of the River, spoiling so much of it for ourselves, but from the River's point of view that is all transitory. Even the great channel dams are only petty, fleeting little restraints. A few miles from where I am writing this, the crumbling Lock and Dam 26 is being replaced by a vast new edifice costing hundreds of millions and which, in the next half-tick of the Mississippi's ancient clock, will, in turn, crumble. No dam can survive such a river's displeasure indefinitely, and it is not the River's pleasure to be blocked and bound. In spite of our contempt for the integrity of great rivers, the Mississippi will shrug off our abuse and move on.

This is the fine delight of rivers. They are born traveling, wanting always to move on, intolerant of restraint and interference— itinerant workers always rambling down the line to see what's around the next bend, growling or singing songs, depending on how things suit them. Now, a lake never goes anywhere or does much. It just sort of lies there, slowly dying in the same bed in which it was born. The lake is a set of more or less predictable conditions—at least, compared to the swiftly changing stream of physical, chemical, and biological variables that constitute a living river. Among those variables, though, is one reliable constant— for me, anyway. Whenever I am out on a river some of its freeness rubs off on me. And since freedom is always a highly perishable commodity, frequent returns to the river are necessary for taking on a new supply.

Most travelers on the Mississippi go with the flow, and the farther they go downstream the more convinced they seem to become that they're really seeing the River and getting to know it. But while going down a river may be the easiest way, it is often seeing the river at its progressive worst. Downstream is where the most commerce exists as a rule, where a river is deepest and strongest, where cities and towns are usually larger and market-places flourish. Downstream takes the traveler into the polluted main line of business where the mature river is so useful in making money and disposing of the wastes that are always a by-product of the money mills.

There are exceptions, of course. At the head of the Upper Mississippi are Minneapolis and Saint Paul, which can match any downstream cities in general pollution and debasement of riparian environments, while down at the far end are Louisiana backwaters and bayou regions as lovely and genuine as riverlands anywhere. I am probably as fond of such places as it is possible for an upstream Yankee to be, but they are just not the same as our upper reaches where the River cuts deep into bedrock. It is country that stands on its hind legs and shows its limestone muscles, rising sublimely over a river that flows in broad running lakes and the tangled multitude of sloughs, cuts, and side channels that wander through a fastness of wooded islands and floodplain forest.

To some of us the most interesting parts of any river are those parts farther upstream, up somewhere back of beyond. Upstream is where the springs are, mainly, and the brightest waters, and most of the rocky outcrops, big timber, and rough ground with waterfalls and rapids and the best fish and fishing. Farther down-stream a river has grown slower and older and heavier, and just isn't as good a place to be young in. That enduring symbol of unfettered boyhood, Huck Finn, was a son of the Upper Missis-sippi. He only happened to head downriver because all he had at the time was an old canoe, and the River was in full flood and really whistling. He might have had more fun if he had headed upstream—which is the choice made in this book. It might have been Huck's, too, if he'd had a good johnboat and outboard motor.

The Mississippi that I know best is the part "between the saints" from Saint Louis to Saint Paul. It is the region of great channel dams, a riverine staircase of twenty-six steps ascending from near the mouth of the Missouri River to the Twin Cities, a series of

great flowing pools reaching for 650 miles in what can be said to be the most manhandled and unnatural stretch of the entire Mississippi, since most of the rest of the River is free-flowing and unchecked by any dams. Yet the lower "free-flowing" Mississippi is largely confined between high levees—a commercial ditch between man-made berms that protect the flat, broad floodplain. Far upstream, the River is contained in a relatively narrow, natural floodplain between sheer headlands, and above and below the channel dams the River splits, eddies, pools, and twists through a bewildering maze of wild islands.

This is the part, from Saint Louis northward to Saint Paul, that some of us call "Upper Mississippi." I've grown accustomed to that designation with years of using a battered little book of river charts published in 1939 by the United States Army Corps of Engineers. This refers to the river above the mouth of the Missouri as "Upper Mississippi." The 195-mile stretch between the mouth of the Missouri and the mouth of the Ohio is "Middle Mississippi." From the Ohio River to the Gulf it's all "Lower Mississippi."

These older designations were eminently logical, but have been revised in a stroke of bureaucratic streamlining that now designates the "Upper Mississippi" as that stretch between the Twin Cities and the mouth of the Ohio, and all the rest as simply "Lower Mississippi." "Middle Mississippi" is done away with entirely. This may expedite paperwork somewhere, but it doesn't make a whole lot of sense out there on the River.

Fact is, there are really four Mississippis.

They begin with the bright little creek that breaks out of Lake Itasca in northwestern Minnesota, flowing north and then bending broadly east and south, running past dark forests of tamarack and spruce, through Indian lands and the beds of wild rice, feeding and draining blue lakes on the way—a clean, free little flowage with the innocence and freshness of youth, mostly unblemished by the corruptions of maturity. This headwaters Mississippi ends at the Twin Cities, where it is separated from the next section of river by a geological and biological division point called Saint Anthony Falls.

It is below these falls that the Upper Mississippi begins, soon fed by the Minnesota and Saint Croix rivers and coming into an older valley that cuts through bedrock and uplands for nearly eight hundred miles. It is gaining strength and maturity as it goes, fed by the Chippewa, Wisconsin, Root, Skunk, Des Moines, and a host of smaller streams, but its midsummer flow back in the pre-

dam days was often too puny to accommodate any but the most shallow-draft commercial vessels.

The River begins to change near Saint Louis, where its volume and silt load are doubled by the Missouri River. The floodplain is widening, headlands are less abrupt, and the wild islands and backwaters have almost vanished. This reach of Mississippi, between the mouth of the Missouri halfway down Missouri's eastern border and the mouth of the Ohio at the southern tip of Illinois, is a transition between Upper and Lower Mississippis. It is not unreasonable to regard that section as "Middle Mississippi," which does not conform to current usage by the United States Coast Guard and Corps of Engineers, but seems to fit the natural state of things and is the system some of us still prefer. This may seem a tad arrogant, perhaps, but it's our river, too—and we've as much right to hang labels on it as the feds have.

Before there can be a river there must be land, and the land must have lift so that snowmelt and rain can run down to the sea.

The highlands that conceived the Mississippi were the tilted floors of ancient oceans, swamps, and forests that had been lifted by deep subterranean convulsions to form a great central plateau. Down through far ages the old seas advanced and receded over germinal landscapes, pressing their basins under vast weights of water. Some were warm oceans, lit by a sun younger than ours, their tides pulled by a stronger moon, the shallows swarming with creatures that have not existed on earth for a half-billion years— myriad little animals with calcareous body parts, thriving in the gentle seas and covering the ancient muds and bedrocks of the seafloors with their remains.

The seas drew back, their beds emerged into the sun and fresh rain to be clothed with mosses and ferns, while our creature ancestors stirred from the sea slime and ventured up onto land. Again, shifts of the deep basement plates and melting of polar ice raised the earth's oceans and sent them across the land. The muds and plant debris of swamps and dry land were drowned by rising seas whose surf pounded shoreline rock to sand, forming a new seafloor. A million years passed, a hundred million. The seas rose, their storm tides and breakers tearing at the coasts as rain fell on the highlands and wore the land from above. Two hundred million years, with the waters rising and then draining away again, leaving beds of sand and calcified animal remains. Swamps emerged from the lowlands with the first tiny reptiles evolving into unimaginable giants, and those passed in their turn to be buried in new muds and sands. Countless life forms, evolving and flourishing and then vanishing forever, fixed in the rock strata commemorating their lost world. Layer on layer the strata thickened and rose—the carbon-rich ooze of ancient swamps and tropical forests, floors of sand and shell from extinct oceans, mudflats from the silty outwash of rivers—each adding its immense weight over deeper layers, compressing and cementing formations of coal, sandstone, limerock, and shale.

A hundred million years ago the existing sea drew away from the northerly plateau, retreating to the lower part of North America. The climate grew gentle and stable, and although a fifth of the continent was still covered by quiet ocean, the land basked in subtropical warmth and forests and deepening soils clothed the ancient rock. The first modern flowering plants appeared, with such trees as sequoia, palm, sassafras, bald cypress, and magnolia. Freed by the disappearance of the great dinosaurs, the advance guard of mammals began to flourish and rapidly evolve. For more millions of years the gentle climate endured, and life bloomed.

Then, through North America's heartland some thirty million years ago, the bland climate began cooling. Palms, figs, magnolias, and other subtropical trees shrank southward, leaving their former range to aspen and conifers and to widening grasslands. It was the cool, temperate preface of the great Ice Age—a continental autumn that saw vast herds of newly developed grazing animals spread across the lovely pasturage in panoramas that man would never know. The gentle epoch was closing.

To the north, an immense store of ice was building and thickening. Centuries of snow fell without melting, deepening and compressing into ice as hard and blue as steel while the world's oceans shrank and revealed land bridges between continents. So greatly thickened that its incalculable weight rendered it plastic and fluid, the ice began to flow slowly southward. With its approach the temperate climate of the northland grew bitter; tall grasses of southerly origin were succeeded by short, wiry grasses, and forests of conifers replaced many of the hardwoods that had flourished during the autumn time.

Beginning a million years ago, the ponderous rivers of ice moved southward in four successive advances, planing and scouring the face of a continent. There were interglacial periods lasting thousands of years when the ice sheets retreated back into their northern strongholds—only to gather weight and strength to advance again. Before it ended, almost a third of the earth's surface would have been under ice as much as a mile thick. Eighteen million square miles would be bulldozed in the world's mightiest back-and-fill operation.

Even before the first ice rivers ground down out of the north, an ancestral Mississippi was becoming one of the master streams of central North America. The continental uplift that had created the Rockies had begun directing drainage eastward, while the older Appalachians were already draining westward into a great embayment that extended far inland from the Gulf of Mexico. One of the main watercourses out of this eastern drainage was a large river that ran in a sweeping curve from the Appalachians, over through what is now Ohio and northern Indiana, and down into central Illinois where it joined the preglacial Mississippi and ran southwesterly along much the same course as today's Illinois River. This was the ancient River Teays, which would be obliterated by a lobe of the third glacial advance. The Mississippi's channel in that region would also be displaced, undergoing repeated shifts until the River left central Illinois for good and took the more westerly course that it still follows.

Of all today's Mississippi River, only the northerly part of the Upper River follows its old preglacial course. The Missouri and Ohio rivers did not exist in their present locations until glaciation had made its most southerly advance, with cliffs of ice that shoved those rivers as far south as Saint Louis and southern Illinois, bringing a climate like Norway's to a region that had grown figs and palms not long before. The headwaters Mississippi that we know today was born after the last glaciation had retreated, draining southward after leaving the gouged and piled wreckage of northern Minnesota. There was no Lower Mississippi as we know it; the

Gulf's embayment reached up to what is now the mouth of the Ohio River and the entire alluvial plain of the Lower River—a thousand miles long and up to seventy miles wide—was largely created by billions of tons of outwash silt, debris, and windblown dust.

As the last glaciers shrank back through what is now the northern Midwest, meltwater pouring from their fronts was prevented by the ice itself from flowing northward. It gathered as vast inland seas of fresh water, largest of which was Glacial Lake Agassiz. Bigger than all the modern Great Lakes combined, this immense puddle of glacial melt was up to 250 miles wide and 700 miles long, covering what is now northwestern Minnesota, eastern North Dakota, much of western Ontario and almost all of Manitoba. Its main outlet was the Glacial River Warren, draining east and south and entering the Upper Mississippi near modern Saint Paul. Above its junction with the River Warren the young Mississippi ran down through raw landscapes only recently vacated by retreating glaciers. There it was a modest sort of river. But below the junction with River Warren it was monstrous.

Draining the largest freshwater lake that ever existed in North America, the Glacial River Warren was an awesome torrent that thundered in from the west, cut deeply into older glacial debris in the Mississippi's channel, and plunged nearly two hundred feet into the preglacial valley below. Here the river ran over a sheet of hard limestone that overlay deep formations of more friable sandstone. As the cauldron at the foot of the cataract ate away the sandstone, undercutting the layer of limestone, slabs of unsupported caprock at the brink of the falls broke away. Progressively undercutting this harder rock, the River Warren Falls slowly retreated upriver. Passing the junction with the Mississippi, the falls continued to move up the River Warren for about two more miles, where the sheet of hard limestone ended. The falls ended with it.

Where the River Warren Falls had passed the Mississippi, the smaller river's mouth was left perched almost two hundred feet above a great gorge. Briefly, for perhaps less than a century, a horseshoe-shaped cataract must have existed at the junction of the two rivers—about as high as today's Niagara Falls and probably

wider—a deafening, mist-shrouded chasm that was surely invested with spirit-powers by the Neolithic hunters who knew it.

The new waterfall of the Mississippi, surviving long after its parent River Warren Falls was extinct, would be the Saint Anthony Falls that marked the biological and geological division between the Upper Mississippi and the headwaters. But from the older waterfall that preceded it, Saint Anthony Falls inherited the mortal weakness that at once creates such places and dooms them. Powerful, churning turbulence in the plunge pools below the falls produces hydraulic plucking and abrasion that gnaws away soft shales and sandstones, with that fatal undercutting of hard caprock at the brink. Saint Anthony Falls were no sooner born than they began their upstream death march, retreating about four feet each year.

For eight thousand years the falls moved upstream, traveling eight miles from their origin and leaving a deep gorge along their line of retreat. They were approaching extinction as they moved upriver to within less than half a mile of Nicollet Island, the place where the hard limestone caprock of the falls would end. Acting somewhat out of character by preserving a major river obstacle instead of removing it, the Corps of Engineers in 1876 built concrete dikes and aprons to protect what was left of the limestone caprock, adding two low dams to keep the limestone under water and protected from freezing and splitting. The falls were, after all, a valuable source of power.

Not long ago my old friend Harold Clemens (no relation to Sam) and I invaded the heart of Minneapolis on a pilgrimage to Saint Anthony Falls. We had wondered why there were no recent photos of it, and why recent authors haven't talked much about it.

We got lost, of course. Clemens and I could get lost in a two-tree orchard, and it is even worse when we get to town. Although asking directions to Saint Anthony Falls evoked a lot of friendliness and fuzzy cooperation, it didn't really help much.

"Well, the falls are down at the river, of course," the young man in a Standard Oil station told me.

"I figured as much. Where, at the river?"

"Right downtown, sort of."

"What's 'sort of'?"

"Well, the falls are there and they ain't. You can see where they are, or where they were, down at the point of the island between the bridges. Right downtown."

"What bridges?"

"Hey, Pete! What'n hell are those two bridges that Saint Anthony Falls lays between? O.K. The Third and Fourth Street bridges, mister. Right downtown."

So we herded my old pickup truck through the morning traffic and across the Fourth Street Bridge to the Saint Anthony side of the Mississippi. The old guy in the parking lot said yes, he'd lived around there most of his life. Saint Anthony Falls? Yep, this is the place. Down that way.

We walked through the spring rain past a sleek new shopping mall being entered by sleek young matrons, and found a path leading to the University of Minnesota's Hennepin Island Hydraulic Laboratory. Just beyond the complex of buildings and chain-link fences was the open upstream end of the island that promised a real on-site exposure to the falls at last.

There were no falls in sight. Not even any rapids. Instead, there was a complex of low dams that sprawled across the Mississippi like a broad, square-topped letter "A." Just below us the river poured over a low-head dam, sliding down its long face into a line of white turbulence. But nowhere could we see any sign of a natural waterfall.

A young man near the lab told us that we could get a much better look at the falls from the other side of the river, probably from the Corps of Engineers buildings at Lock Number 1.

We retrieved the truck and the old guy happily assured us that yes, the other side was the place to go to see the falls. So we drove over there and made all the right turns for once, and pulled into an empty parking lot and were informed by a sign that there was an observation platform at the head of the stairs—the place, ostensibly, from which we would finally feast our hungry eyes on the real, original, one-and-only Saint Anthony Falls of the Upper Mississippi.

The observation platform proved to be an enclosed room with informative exhibits telling us of the engineering history of the lock and dam just below us, and how swift, decisive, and professional action by the corps had conserved the abrupt fall of the River at this point and kept the original falls from retreating up to the point of Nicollet Island where they would have deteriorated into a useless jumble of rapids. There were framed awards from engineering societies attesting to the professional excellence of this.

Of the natural, original falls we could see no sign. But by looking at old photos of the unimproved falls in the exhibit, and then studying the complex of dams and spillways before us, we could see the ledges near the end of Hennepin Island across the river— dry ledges, with slabs of limestone caprock lying at their bases just at the river's edge. It was apparent that no water had flowed over the ledges for years. The rest of natural Saint Anthony Falls had been buried under concrete, water, and money by the corps. Still, as the exhibit pointed out, the natural falls were doomed anyway. And as the framed citations pointed out, it *was* a mighty fine piece of engineering. Anyway, that's just about the upper limit of heavy commercial traffic. The big stuff can pass through the lock into the Minneapolis Upper Harbor for another four miles or so, but that's it.

From the mouth of the Glacial River Warren downstream into the great eastern bulge of modern Iowa, the Mississippi followed a course millions of years old. Ice sheets had never come down this reach of the valley. From Saint Anthony Falls southward, the underlying rock formations rose in a broad arch from below sea level to a crest near modern La Crosse, Wisconsin, where the heights tower six hundred feet over the floodplain. For three hundred miles below the mouth of today's Chippewa River the glaciers split and flowed around this broad upland to leave a driftless area—ten thousand square miles of rugged highlands that were never touched by glaciation of any sort. Through this ran the freshened Mississippi, swollen by glacial melt, roaring down its ancient bed to tear a valley miles wide and hundreds of feet deep. Between headlands and sheer cliffs, this trench of the Upper and Middle Mississippis runs eight hundred miles from Saint Anthony Falls to a point just upstream from the mouth of the Ohio River—a flood canyon carved into bedrock by the great melting a hundred centuries ago, ending at the coastline of an inland sea.

One of my pet fantasies is set in the time when the Mississippi had cut its upper valley to the greatest depth (about two hundred feet deeper than it is today) but was no longer fed by Lake Agassiz, which had finally begun draining northward. There may have been a time when the small tributary streams running from the un-glaciated heights had not yet cut their valleys down to the level of the Mississippi's new, deeper valley. For a while, before the diminished Mississippi had built up its modern floodplain and those tributaries had cut their valleys down to meet it, the little rivers would have entered the main valley in wild and wonderful ways—as long foaming cataracts pouring down rocky channels and over staircases of ledges, or as sheer waterfalls as high as three Niagaras. They'd have been those Minnesota rivers streaming off the west flank of the Wisconsin Arch, ones like the Root, Whitewater, Zumbro, and Cannon—roaring, shouting, spring-fed little torrents as pure and clear as wild water can be, plunging off precipices into the bouldered valley-canyon of a young Mississippi. Damn! I wish I could have seen that!

It may have been even better later on, when the raw lesions left by the Glacial River Warren had healed and the Mississippi flowed more modestly. Before white settlement broke and stripped the watersheds, the Upper Mississippi surely had its turbid times although its average silt load was nothing compared to that of the Missouri River. In midsummer and fall the Upper River ran clean, as it was in the early 1800s when a geologist named Owen walked through a flowering meadow near Prairie du Chien and stood in wonder at the river's edge.

"Imagine a stream a mile in width," he effused in his official report, "whose waters are as transparent as those of a mountain spring, flowing over beds of rock and gravel . . ."

However transparent those waters, they were infinitely more fertile than any mountain spring, for they drained the pulverized mineral wealth of the northern Midwest—the same building stuff that was the base for the fat prairie soils along the watersheds. A rich, clean river moving easily to the sea, its first thousand miles broken occasionally by mild rapids where outcroppings of bedrock disputed its passage, river of giant sturgeon and catfish and paddlefish, of myriad bronze-backed bass and opal-eyed jack salmon, with pearl mussels paving long reaches of its bed. Its sheer cliffs held the eyries of peregrine falcons and twisted cedars from which eagles watched the limitless flights of wildfowl and passenger pigeons. The alluvial terraces of its tributaries were winter quarters for the late woodland people who gathered in summer on the timbered ridges above the River to memorialize their honored dead and celebrate brother creatures by building great effigy mounds depicting falcons, snakes, turtles, and lines of marching bears. Well, it was all worth celebrating—the rich river below, teeming with fish and mussels, beaver, otter, and waterfowl, the upland forests with their deer and turkey and bear, and not far away those smiling sunlit prairie lands with elk and bison and prairie chicken. A region of which the Mesquakies said: "The North is too cold, the West too barren, the East too bloody. This place is just right."

And damn again! I missed seeing that by only two hundred years!

The River took its name, as good places of that kind should, from the red man. Leave it to the Indian—no one can name a place so well. If there are any ugly place names in any Indian tongue, I can't recall them. We usually mispronounce those words and abuse them with our nasal slurring of the vowels and by accenting syllables in all the wrong places. Of any Europeans the French probably do it best, which may be one reason they hit it off so well with Indians. Still, it was a Frenchman's mistake in translating "Mississippi" as "vieux Père des Rivières" that has led generations of Americans into thinking the Indian word means "Father of Waters," an insipid imagery that was never concocted by Indians.

Depending on where you talked to Indians up or down the River, it might be called Sassagoula, Culata, Nomosi-sipu, or any one of countless other names. I wasn't there, but I am told that it was easy for white explorers or even their interpreters to misunderstand local Indians, who might have a different name for each bend, backwater, or bankside feature on their section of the River. Far to the north, people of the Algonquin nation also had several names for the River but some, including the Chippewa and their close relatives, often called it "Mis-sipi" or "Misisipi," "Misi" being a rather broad term signifying "big," while "sipi" was plainly "river." This was to be the most durable name for the Great River, given to Marquette and Joliet by two Miami guides who spoke an Algonquin dialect. It was "Misisipi" to those northern Indians, and the two explorers kept that name as they went on downstream. That is what it has been ever since, with some extra "s's" and "p's" thrown in for emphasis.

If they had started in the south and gone upstream with different guides they might have used a southern name like "Pekitanoui." In fact, their whole way of looking at things might have been different if they had come upstream to the junction of the Missouri and Mississippi.

As that junction was approached from the south the two rivers would appear to be about the same size. There would be good

reason to think that the Missouri was the real continuation of the main river and the Upper Mississippi was only a large tributary; further exploration would verify this. The Missouri River is longer than the Upper Mississippi by nearly sixteen hundred miles and its total drainage basin is three times as large. Even the drainage basin of the Ohio River is bigger than that of the Upper Mississippi. By rights, only that portion of stream between Lake Itasca and the mouth of the Missouri should be called "Mississippi" after the original Indian name. The rest of the mighty river that springs from the Three Forks country in Montana and ends in the Gulf of Mexico below New Orleans should probably be "Missouri River" all the way. After all, the total length of the Mississippi is measured from the Gulf to western Montana, and not from the Gulf to northwestern Minnesota.

But the Upper Mississippi was the first of these two great flow-ages met and recorded by European explorers, and, right or wrong, the name "Mississippi" has stuck. From a strictly semantic point of view, it's no contest. "The Great River" is a far better handle for this majestic north-south system than "Wooden Canoe," which is one way that "Missouri" renders out.

A prerequisite for writing about the Mississippi, it often seems, is to first establish a measure of authority with publisher and reader by traveling the length of the River in some sort of boat. This can be as appealing as it is logical, but there are things about it that have always bothered me.

First, I have never done it and am not likely to. And sitting here on the bank feeling left out while everyone else floats blithely past on their way to New Orleans, I am inclined to kick the nearest cottonwood and tarnish the welkin. Go on. Waste your time floating the whole river, and I hope you get your *deleted* sunburned to a *deleted* crisp! (But I travel light, if you've got some room.)

A notable adventure, there's no denying it. The trouble is, it tends to be a cursory exposure in which many details—if noted at all—may flow together in a panoramic stream that blurs their fine edges—too much riverbank passing too quickly; too much never seen and never fixed into memory. A full-length Mississippi float can inspire a travelogue of the first order but is likely to give priority to the fact of travel and not to the fact of River. The traveler may have floated with four Mississippis through regions as diverse as the planets, and would celebrate the excursion in some French Quarter bistro, still as uncomprehending of the distinctions as when he launched his boat up in Minnesota. Learning those distinctions, and something of the details that compose them, can give you more River. And more fish, too.

Although I've never floated the entire Mississippi I have been on it at one place or another from Lake Itasca to the delta passes. My time on the River has been fragmented into countless visits over the past thirty years—some lasting hours, others continuing for weeks. Almost all this has been from Saint Louis north, for I have never been able to whip up much enthusiasm for the Mississippi below the mouth of the Ohio. A levee-bound channel and floodplain cotton fields may have their charms, but rock palisades crowned with forest and prairie have them beaten hands down.

Little of my river time has been spent on big commercial towboats or sternwheelers trying to recapitulate Mark Twain. I once spent a few days on a towboat—and although that was somewhat instructive it can hardly compare to the years spent in freight canoes and johnboats. Some of my best time on the River has been in the company of game wardens, biologists, commercial fishermen, clammers, trappers, hunters, and a smelly, mud-smeared coterie of river rats in general, and my views of the River are far more likely to reflect theirs than those of the transportation industry.

Which is by way of explaining why this book does not treat of happy "darkies" dancing on cotton bales, steamboats 'round the bend, or the wonders of modern river commerce. I know little of such things; my real interest is in the physical and biological Upper River, in the vital people whose lives are so closely linked to it, and the dangers that beset what is left of the natural Mississippi. That, and the moods of the River—its character and temper in starlight and storm, on August afternoons when water and sun flow together like molten brass, the January days under white skies and a wind of searing cold with the long booming howl of thirty-inch ice expanding across the great river pools, and transition seasons of gentleness and beauty: that special world of mud, scour holes, wild orchids, yorky nuts and fine pearls, of crowfoot bars, trammel nets and heron rookeries. And if, in the examination of these things, I happen to offend certain colonels of Engineers, various farmers, industrialists, and proponents of unlimited barge traffic, I can only say that the offense is offered wholeheartedly and in a devout spirit of truculence.

1

THE OLD RIVER

1

In *terra incognita*, if the opportunity presents itself at all, the only way to go is by river—always assuming, of course, that you and the river happen to have the same general route in mind and that the river doesn't object violently to having passengers. At the same time, there is a certain comfort in knowing where the thing ends and where it begins.

De Soto found the main trunk of the Lower Mississippi in 1541. La Salle sailed down it to the sea in 1682. Marquette and Joliet came into the middle reaches of the Upper River in 1673. Sieur Du Luth visited the headwaters region in 1679; Père Hennepin was there in 1680. By 1806 the whole Upper River was American territory and the army had built Fort Snelling below Saint Anthony Falls, and *still* the true source of the Mississippi River was unknown.

There was a rumor that it came out of a hole in the ground up around Hudson Bay or somewhere, but no one had ever found it. Oh, they had tried. Sieur Du Charleville gave it a shot as early as 1700, coming from far down the Mississippi up to a point just below Saint Anthony Falls. There the Indians told him it was about as far upstream to the River's source as it was downstream to the mouth, and that did it for Du Charleville. He went back the way he had come, without the slightest desire to continue up a river that apparently began somewhere in the Arctic.

When Lieutenant Zebulon Pike headed upriver from Saint Louis in 1805 he did so under orders to "proceed up the Mississippi with all possible diligence," select sites for future military posts, appreciate the Indian situation, and "ascend the main branch to its source." It was a slow trip upstream and winter caught them behind schedule. They made winter camp near what is now Little Falls—about a hundred miles up into the headwaters region from Saint Anthony Falls—and continued on foot, traveling on the river ice by sled. In early January they had gone another hundred river miles to the Northwest Fur Company's post at Big Sandy Lake, where they rested for a couple of weeks before pushing on.

They soon found the Mississippi bending into the northwest in a sweeping arc, passing Lake Pokegama and the Vermillion River. Not far beyond little White Oak Lake, ill equipped for the deep cold through which they were traveling, they came to a fork in the River. Pike chose the south branch, a strategic error that led him up the Leech Lake River and away from the fledgling Mississippi. He reached Leech Lake in the bitter evening of February 1, 1806, planting the flag on the shore and proclaiming the lake to be the "main source of the Mississippi." One version of this adventure has the Indians telling Pike that this was indeed the main source and he, not really believing them, striking overland to the northwest and Upper Red Cedar Lake, which he proclaimed to be the "upper source" of the Mississippi. But like several subsequent explorers, the young lieutenant hadn't gone far enough, nor west enough.

At the time, Pike's discovery may have been accepted as valid, but before long there was growing interest in verifying his claims and carrying his explorations farther. However, it was fourteen years before the next real attempt was made, this time by the territorial governor of Michigan, Lewis Cass, who led a thirty-eight-man expedition from the southern end of Lake Superior up into the northwestern Minnesota wilderness. In July 1820, the party reached the Upper Red Cedar Lake where Pike had halted. In honor of their leader, the men voted to name this lake "Cassina," which was later trimmed to "Cass Lake," which it bears today. This was also as far as Cass's party went, but they did not really believe they had found the Mississippi's source. Their topographer, David Douglas, wrote that the Mississippi actually emptied into the west side of the lake and that the little river "took its rise in a small lake called 'Lac la Biche'." The official journalist of the party, James Doty, recorded that "they did not go to the extreme source of the river, only to red cedar Lake."

Next to try was an expatriate Italian, one Giacomo Constantino Beltrami, whose burning ambition was to be numbered among North America's great explorers—and figured that North was as good a compass point as any at which to begin conquering.

In 1823 he arrived in the United States and hurried down the Ohio River and up the Mississippi to Fort Snelling at the mouth of the Minnesota. Major Stephen H. Long was about to resume the army's northern explorations and was preparing to lead a party up into Canada and the Lake Winnipeg area. Beltrami persuaded the major to let him accompany the expedition part of the way; at Pembina, on the Canadian border, he left the main party and struck off southeastward toward the unknown source of the Mississippi. He wanted adventure, and found it. His Chippewa guides deserted him in unexplored country but he somehow won through to the Red Lake region. Just south of there, between Red Lake and Cass Lake, he came upon a little heart-shaped lake that he named "Julia" and which he proclaimed was the true source of the Mississippi.

A bit too flowery and romanticized for the pragmatic tastes of American military and scientific bodies, Beltrami's two-volume account of his expedition and discoveries was unconvincing. But while he may have failed to persuade authorities that Lake Julia was the head of the River, he fueled new interest in further exploration.

This was not hard to do.

The exploring business was a matter of great public excitement about then, what with the young nation caught up in the fever of Manifest Destiny, and all that new ground in the Louisiana Purchase waiting to be inventoried, and it was as good a way as any to become an overnight hero. Moreover, a famous river was involved—and if there is one thing that fiddle-footed citizens of all times and places can never resist, it is a river with an unknown source. Of course, the Indians didn't seem very excited about it all, and for good reason. Some tribes had lived around there for several thousand years and at one time or another had camped and hunted all around the source, and if there had been anything really remarkable about the place, they'd have known it. Otherwise, one creek in their backyard was pretty much like any other. But then, the Indians were not mapmakers and took no honor in publishing learned reports. To certain Europeans, on the other hand, the urge to fill in the blank places on maps was a fever in the blood, a spur and an itch that would always drive them back of beyond until the blanks were filled—usually with the names of wives, politicians, or the explorers claiming to be there first.

For all practical purposes, the search ended in 1832.

Henry Rowe Schoolcraft had left his native New York state in 1817 and quickly racked up an impressive record on the northwest frontier. On the strength of a book he had written of a visit to the Missouri lead mines, he was invited to join the Cass expedition as mineralogist. This resulted in a second book, his *Narrative Journal of Travels through the Northwestern Regions of the United States to the Sources of the Mississippi River.* Published in 1823, this surely had much to do with Signor Beltrami's decision to go adventuring.

Schoolcraft married a well-educated half-Indian woman named Jane Johnston, and with her help he quickly became an authority on the customs and politics of the Chippewan people. He served in the Michigan Territorial Legislature from 1828 to 1832, was an Indian agent, and became a cultural leader of the territory. It was his writings about the Chippewa, by the way, that inspired Longfellow to create *The Song of Hiawatha.*

His interest in the northwest territory had never subsided, and the deep-grained itch to find the source of the Mississippi once and for all had not diminished in the decade since the Cass expedition. But he was a full-time Indian agent with little money and even less reason to go looking for the top end of a river. Besides, the only sensible time to do so would be during the summer months, and that was the busiest season at the Sault Sainte Marie Indian Agency. It seemed a hopeless dream unless he could quit his post and finance an expedition with his own purse, and that was out of the question.

Sometimes, though, obstacles have a way of becoming bridges. When Schoolcraft's main chance to find the River's source did come, it was because of Indian problems and not in spite of them.

For almost three hundred years there had been unremitting hostility between the region's Sioux and Chippewa. The latter were relative newcomers to Wisconsin and Minnesota, having been driven westward by the powerful eastern Iroquois. Although the Chippewa were no match for the warriors of the Iroquois federation, they were strong fighters who moved into the Sioux country with guns, while the early Sioux were still warring with arrow and club.

The bloody raids and ambushes showed no sign of abating in the late 1820s, when Sioux twice slaughtered Chippewa in the shadow of Fort Snelling's palisades, and the federal government grew seriously concerned for the first time. Up until then, the only whites who might be caught in this intertribal warfare were a few trappers, traders, adventurers, and the usual offscourings of civilization who were always found along American frontiers. If such were unable to protect themselves, there was no great loss. But all this was changing. Settlers and their families were arriving in growing numbers—the sort of people who create a solid economic base and stable political systems—and the government had a vested interest in their health, welfare, and votes. In a crash program to quiet the frontier, treaty conferences between the warring factions were organized, Indian leaders were pressured to make pledges, new forts were built, and Henry Schoolcraft labored tirelessly to "check and allay the spirit of predatory warfare" among the Chippewa whom he knew so well.

One of his urgent recommendations was to establish a sort of territorial boundary between Sioux and Chippewa that each tribe would recognize and respect. He also suggested an expedition into the northern Minnesota country to further "check the predatory spirit" of the Chippewa there, the sort of effort that had already succeeded in Wisconsin. Up to this point, nothing in any of his letters and recommendations revealed an intention to look for the source of the Mississippi. If he had any secret plans for doing so, however, they weren't weakened by the fact that Lewis Cass now happened to be secretary of war. Since the Indian Office was then in the War Department, there was never much doubt that School-craft's expedition plan would be approved. Shortly after that, he wrote Cass: "If I do not see the 'veritable source' of the Mississippi, this time, it will not be from a want of intention." On June 7, 1832, the expedition of five "gentlemen" members, twenty boatmen and guides, and an escort of ten soldiers left Sault Sainte Marie and headed into the northwest.

Traveling past Cass Lake, the party arrived at Lac La Biche on July 13. Lieutenant James Allen of the military escort noted in his journal that "There can be no doubt but that this is the true source and fountain of the longest and largest branch of the Mississippi." Schoolcraft, as might be expected, was somewhat more exalted by the occasion: "We followed our guide down the sides of the last elevation, with the expectation of momentarily reaching the goal of our journey. What had been long sought, at last appeared sud-denly. On turning out of a thicket, into a small weedy opening, the cheering sight of a transparent body of water burst upon our view. It was Itasca Lake—the source of the Mississippi." They didn't stay long. After about four hours at "Itasca" they turned their backs on it and headed back downstream to Cass Lake. And that was that.

The headwaters lake was "Lac La Biche" in French and "Omushkos" in Chippewa—both of which rendered out as "Elk Lake," since the lake bore a rough resemblance to the head of a wapiti. Yet, Schoolcraft renamed it "Itasca." Vanity, and pride of discovery? Not entirely. A good deal of confusion attended this name change, but it would seem that Schoolcraft's main reason was based on the fact that "Elk Lake" gave no indication of the place's significance as the source of the Mississippi. The word "Itasca" did. It is not of Indian origin as it is often thought to be, but is a coined word formed of the last and first syllables of the Latin *veritas caput*, meaning "true source." Which isn't bad, at that.

Schoolcraft's claim of finding the true source of the Mississippi was nailed down a few years later by a skilled scientist-astronomer, Joseph Nicollet, who carefully surveyed the Itasca basin and precisely established its latitude, longitude, and elevation. He described five creeks flowing into Itasca and described the largest of these as "truly the infant Mississippi" but did not challenge Schoolcraft's discovery. His own further work, Nicollet reported, was simply a more refined version of Schoolcraft's findings.

The search for the River's source had been dramatic enough and had fired the public imagination from the first. A classic adventure in the best frontier tradition—and to some men, it was just too good to be over. So in 1881 a latter-day Beltrami named Willard Glazier announced his intention of discovering the *real* source of the Mississippi.

Glazier was an ex-captain of the Union Army and a veteran of the Civil War, the "soldier-author" who later wrought such disparate works as *Peculiarities of American Cities, Ocean to Ocean on Horseback,* and *Down the Great River.* It was his intention to travel the Mississippi from end to end and was apparently convinced (or at least hoped to convince others) that Schoolcraft had neglected to "coast Itasca for its feeders, and thus missed the goal he had so industriously sought." Further, he contended that even the keen Nicollet had overlooked the main stream entering the southwestern arm of Lake Itasca, and "to have accepted conclusively the statements of those who had preceded him." He resolved to set things straight and personally reveal the primal reservoir of the Great River.

During June 1881, Glazier was in Saint Paul outfitting for the six-week expedition into what was then still wilderness. He was joined by his brother George and Barrett Channing Paine, of Indianapolis, and the three intrepid explorers departed the city on July 4, heading for the Mississippi's source via Leech Lake and the Kabekona River.

An engraving in his book *Down the Great River* shows Glazier standing in a bark canoe, rifle in hand, waving his hat in farewell to an assemblage of polite Chippewa standing on the lakeshore. The canoe, which appears to be about twelve feet long, is impelled by a pair of dusky paddlers. (Glazier, who could not swim, had already entertained the Indians by stepping into a canoe and tipping over in shallow water.) There are two other canoes bearing George Glazier and Barrett Paine, each paddled by an Indian. With three birchbark canoes and four faithful guides, the trio set out on their great adventure.

By the time Glazier and his companions arrived at Itasca they had lost all their fishhooks and were nearly out of ammunition, since they persisted in shooting at targets along the way. They had originally planned to live off the land, but managed to shoot only one duck. When they finally reached the lake their provisions had been reduced to a bit of bacon and a few pounds of flour. The disgusted Indians were all for heading back, but our adventurers pressed stoutly onward and forced their way through a marsh and into the small creek entering the southwestern arm of Itasca.

Struggling up this little creek, they finally broke out into a lake about a mile in diameter. They found this to have three small feeder streams, two of which were spring runs while the third was the outlet of a small lake about a mile beyond. Glazier named this "Lake Alice" in honor of his daughter, but for some reason did not regard it as the true source of the Mississippi. That honor was reserved for the lake they had first found, where Glazier lined everyone up on a small promontory and gave a speech about having "corrected a geographical error of a half century's standing." As he concluded, volleys were fired for each member of the party—which just about did in their remaining ammunition and confirms the suspicion that they were mighty slow learners. Then, to the leader's manifest "surprise," the place was named "Lake Glazier" by unanimous decision. Glazier demurred, of course, but allowed his protests to be overridden by majority opinion. Then, mosquito-bitten and famished, they carried their great news downstream to a waiting world.

Although Glazier's claims were supported by a few newspapers, politicians, and, of course, the publisher of his book, they were never officially recognized. In 1889 the District of Minnesota mounted a thorough topographic and hydrographic survey of the Itasca Basin. The survey lasted two months and resulted in official denial of Glazier's claim to having found a "truer" source of the Mississippi. Not even Glazier's place names have endured. Lake Glazier inherited Itasca's original name and has been Elk Lake ever since. Lake Alice was renamed "Little Elk Lake."

All this must have seemed a rather silly little exercise to any Indians involved in it, especially one like Chenowagesic, chief guide of the Glazier party. He had farmed and hunted around Lake Itasca and its environs for years, and Glazier's "discovery" might be compared to Chenowagesic's discovering Minneapolis. Again, whether or not a particular lake was the true head of the Mississippi was a matter of supreme unimportance to the local Indians. Not that the concept of a river's source was too abstract for them, but that was a geographer's goal, and the Chippewa carried perfectly good geography in their heads. Any idea of achieving instant glory by visiting the head of some insignificant creek never occurred to them—and why should it?

In his superb chronicle of a vagabond snapping turtle, *Minn of the Mississippi*, author Holling Clancy Holling has an Indian boy saying:

> All these lakes an' swamps start th' Mississippi River rollin' —like a sponge leakin' water. My people, the Chippewa Indians, always hunted an' fished an' picked berries 'round here. Then white men began comin'—huntin' for th' HEADWATERS OF TH' MISSISSIPPI! When Mr. Schoolcraft came to Lake Itasca, a hundred years back, he named it from words meanin' "TRUE HEAD" of th' Big River. But other men still talked an' argued. Whichever water was highest *above* Lake Itasca, they said was th' *real* head!
>
> My people made little jokes around their campfires. They had watched surveyors, all wet an' muddy up to their beards, measurin' swamps which *might* be higher'n th' one before. My Great-grandfather, young then, he guided some surveyors. One rainy day he runs into a surveyor's tent an' yells, "You say ELK LAKE is higher'n LAKE ITASCA? Me, I find *LITTLE* ELK LAKE higher'n ELK! No, don't

go measure it yet! Me, I find *pond* higher'n LITTLE ELK LAKE! No, don't go. I find *little spring* higher'n pond! Hey, wait! I find *big tree* growin' over spring. Way up on top, I see CROW a-sittin'! *NOW*—EVER'BODY RUN QUICK AN' MEASURE WHICH END OF CROW IS HIGHER'N OTHER END! AN' *THAT* IS TRUE BEGINNIN' OF MES-SIPI, GREAT FATHER OF WATERS!"

Lo, the wise Indian.

Today, the "true head" of the Mississippi River at the outlet of Lake Itasca is bridged with stepping stones and tourists in season, and a carved post proclaims it to be the source of the Great River that flows southward for 2,552 miles to the Gulf of Mexico. In the shank of the tourist season the fusillade of camera shutters is deafening; approximately every third person in the human race wants to be photographed standing on a rock at the extreme upper end of the Great River.

We had been working upriver all the way from Alton, Illinois, about thirteen hundred miles downstream, trailering the big johnboat and cartopping a canoe. I had hoped to launch the canoe in the very outlet of Lake Itasca, which might have been done easily enough, but a bridge culvert not far downstream would have blocked us. We put in just below that bridge, about a hundred yards short of being able to say we had started in the exact source itself. We wouldn't be going far—less than four miles to the first take-out—but we wanted to get a feel of the River's starting.

It was a bright, gentle little stream no more than twenty feet wide in most places but easily deep enough to carry the canoe with my daughter Josie and me. It seemed in no particular hurry to go south and grow old, heavy, and majestic—an infant river that purled along happily, chuckling to itself now and then as it played around a rootwad or past a log drift. We were still in Itasca State Park, and above the willows and alders that crowded the stream's edges we could see the lofty crowns of white pines that were already a half-century old when Schoolcraft first passed them—the kind of trees called "bull pines" by some timberjacks. (I once asked an old Swede woodsman why this was so, and he roared with laughter. "Vy? Because dey ain't never been cut yet, dat's vy!") We loafed

along over the glass-clear water, our canoe shadow running over the streambed beneath us, and I remembered that Swede and what he'd said and considered the propriety of a great river's being born in an original basin with its grove of bull pines instead of a gelded forest. The River would be abused, strictured, and poisoned aplenty as it matured, God knew, but here and now, for a little way, it could revel in running young and clean through its splendid nursery.

We were coasting on the gentle current, touching the water with a paddle only now and then if the bow swung askew. There is nothing more sneaky than a drifting, olive-drab canoe with motionless passengers wearing neutral, weathered clothing, and we were within thirty feet of the large buck deer when he sprang to his feet. He had been bedded in stunted willows, his sleek red

summer coat blending perfectly with the stems. He made two prodigious leaps through the hock-deep shallows and stood watching. His antlers were almost fully grown within their sheath of velvet, broadly arching and heavy, testimony to a good supply of rich browse. Then he'd seen enough, and two more airy leaps took him out of sight in the heavy alders.

The distant pines drew back even farther. The little river widened, slowed, became marshy with edges of bulrush, lotus, cattails, and wild rice. The channel grew confused, not knowing where to go. Nor did we, becoming as confused as the channel and wandering up blind alleys and bays through stands of bulrush and broad flats of coontail. Here the little Mississippi was more marsh than river, just as it is a bit farther down where it passes through chains of lakes. The River begins in marshes—and ends in them.

The place was alive with Brewer's blackbirds and redwings;
mallard hens flushed from the sedge flats just ahead of us, and
twice blue-winged teal went through the crippled-mother act to
lead us away from hiding broods. An old pine log at the head of
one blind alley was strewn with otter and mink scats, and twice
that day we had seen ospreys. It would have been a good place
without seeing such stuff, but it is always heartening to have an
opinion confirmed by unimpeachable authorities—and deer, os-
prey, otter, beaver, waterfowl, and beds of wild rice endorse the
quality of a place just by being there. And glory be, no deerflies.
Two days before, canoeing across little Squaw Lake, we had been
assailed by clouds of them. That will get our attention every time,
and Tar made it even more interesting. If you're ever jaded with
the humdrum course of ordinary affairs and need a stimulating
diversion, take a small canoe across a northern lake with a seventy-
five-pound black Labrador retriever that's being driven bonkers
by deerflies.

Today, though, no deerflies or mosquitoes, and we were a
month late for blackflies. A fine day in a fine place, with the best
of company, watching a river being born.

A couple of summers later I was thinking of all this, and of the
awestruck tourists crowding around the little outlet of Itasca Lake.
I was in southwestern Montana at the Red Rocks National Wildlife
Refuge, eavesdropping on nesting trumpeter swans and relishing
the most remote refuge in the contiguous states—and possibly the
most scenic. The marshy valley is ringed with peaks. To the west
lie the Lima Peaks that are part of the Bitterroots. Just to the north
are the low Gravelly Mountains, while off to the east in the direction
of Yellowstone are the Hillgard Mountains. Closest of all is the
Centennial Range, a massive barrier rearing up out of the south
edge of the refuge. I was asking Barry Reiswig, the refuge manager,
about the Centennials. "Fine country up there," Barry noted. "You
know, those mountains have a special distinction. They say that's
where the Missouri River actually rises. So I've heard, anyway—
that the highest and farthest of all the Missouri's feeders begins
up there."

Well, I thought. Yes indeed, and well. So that's where the Big
Muddy springs. And depending on who's adding up river mileage

and labeling the map, this just might be the place where the Mississippi begins. There's much to be said for the contention that a "tributary" over twice as long as the entire Upper Mississippi just may be the main stem itself. In that case some name-changing would be in order, since "Mis-sipi" is a Chippewan word that doesn't apply to a river springing in the Rockies. So would it be "Mississippi" from Lake Itasca to Saint Louis, and "Missouri River" from there to the Gulf? And how would the good State of Mississippi feel about changing its name? It's all too much bother, of course.

Still, logic is logic, and on the roof of the Centennials in some alpine meadow a snowbank has issued a little rivulet that's gone searching for the southern sea. I've got to go up and check that. There can be no question about *that* beginning. Not up there above timberline with no treetops where a crow can perch in the rain and drip the beginning of the Great River.

2

The canoe, fashioned either of bark or hewn from a log, was the first workboat on the Upper Mississippi.

It came in all sizes, from the trim little one-man hunting canoe so cunningly fashioned from the bark of paper birch, spruce roots, animal fat, and resins, to the big freight canoes that brought early French holy men and soldiers over the Great Lakes, across on the Fox or down the Illinois, and into the River itself. The canoe still served for personal transport in the Upper River and its branches far into the period of American settlement. Made of common materials, easily repaired, light to carry but tolerant of remarkable loads, maneuverable in wild water and utterly silent when the needs of hunting or war demanded, the *canot du nord* was a remarkable contribution to wilderness commerce. It is doubtful that any craft meant for use in inland waters could have proven so versatile. But not even the best of canoes could meet the rising demand for craft that could handle the heaviest river freight. As that need was developing, however, so was the boat that could fill it.

No one is sure where the prototype of the American keelboat emerged; it may have been at the head of the Ohio, for one of the centers of keelboat-building was long in Pittsburgh. According to biologist Tom Waters in his fine book *Streams and Rivers of Minnesota,* it was first used on the lower Ohio and Mississippi, and by the French on the Upper Mississippi as early as 1751. It wasn't until later, though, that the keelboat was put to its greatest use by Americans exploring and consolidating their vast new acquisitions in the Louisiana Purchase. Lewis and Clark's Corps of Discovery took keelboats up the Missouri, and at about the same time Lieutenant Zebulon Pike was going by keelboat to Saint Anthony Falls. Soon after that, the Saint Louis fur companies were sending their mountain men up the wide Missouri in keelboats, while on the Ohio and lower Mississippi a hyperactive keelboatman named Mike Fink was earning his niche in American folklore by gouging eyes and biting off ears.

This keelboat was a remarkable thing—one of those unique American tools, like the curved-helve axe and the Lancaster "long rifle," that were developed by ingenious people who had found that Old World tools didn't really serve their needs. It was built to move great quantities of freight up and down the western rivers as efficiently as possible, a boat from fifty to eighty feet long, pointed at both ends, with a beam of fourteen feet or more. There was a depth of hull of about four feet. A stout keel ran from bow to stern, the strong ribs sheathed with caulked planking. The boat was usually fitted with a cargo box that filled most of the hull and rose several feet above the gunwales. All in all, it was a remarkably effective and versatile craft. A keelboat might carry a crew of ten men with as much as three thousand pounds of freight per man, and do it in as little as thirty inches of water or less, against a current running six miles an hour.

The boats were driven by wind or current or manpower, and sometimes by all three. For downstream work in favorable wind, sail and rudder were enough. With luck, there might even be a fair wind for upstream travel as well; if not, five or six pairs of large oars might be put into play. Those were all the easy ways. If oars and sail didn't fit the situation, the men would go ashore and sweat the boat upstream with *la cordelle,* a thousand-foot towrope handled by men because the wild shorelines were usually too rough for draught animals. And if some shoreline feature such as a sheer cliff kept even men from going ashore, they would take up the poles.

These were usually made in Saint Louis, long poles of turned white ash with a wooden "shoe" at one end for poling over soft mud bottoms, and a round knob at the other. Running along each gunwale on either side of the cargo box was a narrow wooden catwalk, the *passe avant,* to which were nailed wooden cleats. When poling became necessary, crewmen would line up on these catwalks, facing the stern. The captain would shout: "*À bas les perches!*" and the crewmen would plant the poles in the riverbed, the knobbed ends of the poles in the hollows of their shoulders, and walk toward the stern, slowly pushing the keelboat forward. When the man farthest aft had gone as far as he could, the *patron* would bellow: "*Levez les perches!*" and the men would raise their poles, walk quickly back toward the bow, and again obey the command "Down with the poles." If there was a headwind, or the current was particularly strong and the cargo heavy, the keelboatmen might lean so mightily into the poles that they could catch the cleats of the *passe avant* with their hands and would strain along the narrow catwalk on all fours.

The men who cheerfully submitted to such herculean labors were a special breed, often from the Saint Louis region, men of French blood with a dash of Spanish. These Mississippi and Mis-

souri River keelboatmen were generally regarded as being better at the job than typical French Canadians—possibly because many of them were second- and third-generation rivermen. Contemporary travelers were always struck by the essential toughness and resilience of these Saint Louis Creoles, for after a dawn-to-dark day of grueling labor and a meal consisting of little more than salt pork, hominy, and beans, they would break out a fiddle and dance to the old French reels that can still be heard around places like Bon Terre, Missouri, and other parts of the Missouri Lead Belt. During the brief, dramatic heyday of the fur trade and keelboat, they were undisputed masters of the River, with the special touch of Gallic *panache* that always seemed to mark the frontier *voyageur*. They and their keelboats were pioneers that vanished with the coming of steam.

The first steamboat to get up the Mississippi into the Minnesota country was the *Virginia*, which arrived at Fort Snelling in 1823. And although the keelboat would be a common freight handler into the 1830s, there suddenly seemed to be steamboats everywhere. They materialized as if by magic—a swelling traffic of side-wheelers and stern-wheelers that ran from Saint Louis to the head of navigation at Saint Anthony Falls (where some were taken apart, to move above the falls, and ply the headwaters region), bringing visitors, settlers, freight, tools, and trade goods, and carrying back furs and produce. There were regularly scheduled packets that hauled freight and passengers both ways, and with the coming of the steamboats northern development began in earnest.

There is no part of the Mississippi throughout its whole navigable length that isn't replete with hazards: snags, floating logs, shifting channels, sandbars, low water, high water—any of these things can put you out of the boating business in short order. In addition to all that, the Upper River had some special hazards: the long stretches of rocky rapids largely missing from the Middle Mississippi and entirely lacking in the Lower River. The lower chain of rapids ran for about nine miles from Keokuk to Meltross, while the upper rapids—which were somewhat more dangerous— were fifteen miles of rocky ledges, outcrops, and boulders that extended from Le Claire, Iowa, to Rock Island, Illinois. Not that these were foaming cataracts; they were anything but that. But there were long stretches of river where the bedrock ledges and great jagged slabs lay just beneath the surface waiting to tear the bottoms out of unwary steamboats. There were other "rapids" in the Upper River but these were the worst.

For six years, from 1817 to 1823, Major Stephen H. Long of the United States Army Corps of Engineers explored the Upper River, looking for ways it could be improved for the settlement and commerce that were steadily growing. He recommended, among other things, that canals be constructed around the rapids so that larger boats could navigate the river. It must have been an impressive report, for Congress summarily assigned the responsibility for managing the River and improving it for boat traffic to the Corps of Engineers of the army, where the authority has rested ever since.

Working the River in a flatboat or a keelboat could be touchy enough, but when steamboats came into the picture, they brought their own problems. With heavier cargoes and passenger complements, and competitive pressures demanding that they be driven under power at higher speeds both upstream and down, the typical paddle wheeler not only had more to lose but tended to lose it more easily on that thinly diluted obstacle course of rocks, ledges, rapids, sandbars, and snags. Of all these, snags were the most pestiferous—the dreaded "hull inspectors" that ambushed luckless boats of every kind, and steamboats in particular. There were the "planters," whole trees that had become solidly embedded in the river bottom and anchored there with tons of silt. "Sawyers" were logs also embedded in the river bottom, but with one end free and pointing upstream, the current imparting a bobbing motion as if it were "sawing." A "sleeping sawyer" was hidden just below the surface. If such snags happened to be heavy logs, they could easily punch through the hull of a paddle wheeler.

For thousands of years such debris had been building up in the Mississippi, and the River's ancient logs, stumps, and sunken trees proved deadly to the newfangled steamboats that had already had enough to worry about, what with exploding boilers and all. Before 1849 nearly 30 percent of all steamboats were wrecked in accidents—and of those, over 40 percent were done in by snags of some kind. Little had been done by 1824 to solve the problem;

indeed, little could be done, but in 1829 the first really successful government snag boat was developed by Captain Henry Shreve. It was a clumsy, twin-hulled monster named the *Heliopolis,* as ugly as any boat could be with its great iron-sheathed snag beam connecting the two hulls—but it worked. This mechanical behemoth was driven like a battering ram into a protruding snag. Something had to give, and always did, and it was never the *Heliopolis.* The snag would be torn up, dislodged, or broken off, no matter how large or how solidly anchored in sand and silt. The iron beam could easily tear out a snag weighing seventy-five tons and buried deep in the riverbed; it lifted out one waterlogged tree that was a hundred sixty feet long and three and a half feet in diameter. Such snag boats, known as "Uncle Sam's tooth pullers," had an almost immediate effect on the middle and lower parts of the River. By 1830 the worst obstructions between Saint Louis and New Orleans had been largely cleared away and attention was turned to the Upper River—although it would not be until 1867 that the ancient snags of the Upper Mississippi would be entirely removed.

The problem of the Upper River's rapids wasn't so readily solved. In 1837 two young army lieutenants, recent graduates of West Point, were sent to the Upper Mississippi to survey the Rock Island and Des Moines rapids. One was Lieutenant Montgomery C. Meigs; the other was Lieutenant Robert E. Lee. Under Lee's direction many of the larger rocks were removed from the channel, and that helped, but the Rock Island rapids remained a major obstacle to heavy river traffic until the Moline Lock was built at the east end of Rock Island and boats were able to bypass the worst of it. That lock would not be finished until 1907; six years later the teeth were pulled out of the Des Moines rapids with the completion of the Keokuk Lock—part of a privately built power dam intended to generate electricity. Even with those locks, however, heavily laden boats still drew more water than parts of the river had to offer. At low water in midsummer, when it was possible to wade across the Upper Mississippi, the problem was critical. At such times, certain captains claimed, a steamboat would actually raise a dust cloud. Someone suggested that a railroad be built right up the middle of the riverbed, but that idea was discarded on the grounds that there wouldn't be enough water for the boilers of the locomotives.

However, in spite of snags, rapids, boiler explosions, or low water, the paddle wheelers plied the Upper River. From the 1820s through the 1840s steamboat activity on the Upper Mississippi was actually greater than on the Lower River, although tonnage was less because of rapids and such barriers as the great sandbars at the mouth of the Chippewa River. It took some doing, because there just wasn't as much water up there as there was below Saint Louis, but there were captains equal to the job.

In *Life on the Mississippi*, Mark Twain has Cap'n Bob Styles "reeling off his tranquil spool of lies," including one particularly outrageous yarn about driving a steamboat twenty-three miles without a rudder. Twain admired this embroidery as being worthy of a great liar, and left it to posterity on that basis. But the only lying involved was that Bob Styles did it, and that it was twenty-three miles. The real thing was managed by Captain Bill Tibbals when he was chief pilot on the old *Key City* and drove without a rudder down forty miles of Upper Mississippi.

The *Key City* was heading downstream in heavy fog and was just above Trempealeau, Wisconsin, when she struck something that tore off her rudder. (All steamboats on the River, by the way, were "she," whatever their names. The *Robert E. Lee* was "she," and the *Delta King* was the sister ship of the *Delta Queen*.) The crew managed to fish out the rudder but couldn't replace it because the irons had been broken out.

The Old Man turned to Tibbals and asked: "Can you run her without a rudder?"

"Damn right I can," said Bill. "Start her out and let's go." He knew he could, too, because this wasn't one of your common stern-wheelers that would have been mortally crippled with a broken monkey rudder. This was the *Key City* herself, pride of the River, as agile as a cat and smart as paint, a side-wheeler with paddles driven by independent engines that could be controlled separately. So Bill just backed the *City* out from the Trempealeau landing and headed downstream as cool as you please. It wasn't long before they saw the chimneys of the *Northern Belle*, one of the fastest packets on the River, and passed her easily. Tibbals threaded the big steamboat downriver through two twisty sloughs and made several crossings to stay in the winding channel, using the speaking tube instead of the engine bells, yelling down to the engineer: "More on your labboard! Stop the stabboard! Go ahead on it! Now ease off that labboard!" and so on. They docked neatly at La Crosse, one of the trickiest landings on the River, coming in below the landing and rounding-to in a wide semicircle and landing with the bow upstream. That call completed, the *Key City* continued downstream, "patting her feet nicely," as they used to say, finally putting in at Brownsville, Minnesota, where the rudder was remounted. Forty river miles had been navigated with "no more rudder than there is on a church."

Much of the credit for this epic feat has to go to the old *Key City,* of course. Tibbals gallantly claimed that she was "the grandest steamboat that ever climbed a bar." Another old riverman remembered that "You could thread a needle with the old *City,*" and his friend West Rambo agreed, adding "Why, you could move her wheel with your eyelash!"

Such a boat had the stuff of legends built into her at the ways—but it remained for a cunning old riverman like Bill Tibbals to draw it out, swearing steadily and lovingly as he drove his rudderless vessel into the fog and beyond the works of ordinary boats and men.

'Way for a lightning pilot!

3

The upper watersheds of the Mississippi, Chippewa, Saint Croix, and other large northern rivers were cloaked with great stands of white pine. Few timber resources have been so ideally situated in conjunction with a ready market. Through the late 1830s, and into the 1840s and 1850s, settlers were pouring into the Grand Prairie of northern Illinois and across the River into Iowa. The first of these pioneers had plenty of trees for constructing cabins and barns, but as they moved farther out into the treeless prairies and plains the need for building materials became acute. For a time, finished lumber was even being brought down the Ohio and then up the Mississippi—some of it from as far away as the westerly slopes of the Alleghenies and all of it at much expense.

Yet, stretching across the northern Midwest through Michigan, Wisconsin, and most of Minnesota was the greatest pinery in the world, a forest the size of New England up there at the heads of big easygoing rivers that seemed designed by Providence for the sole purpose of floating logs. All that were needed were the men and means, and these were not long in coming.

There ensued one of the most rapacious romps ever taken through an American resource. The northern wilderness at that time was all public land, and nearly inaccessible, but lumbermen began to break into it where they could and cut what they were able, public land or not. No one cared.

The cutting would be legalized—or, at least, some of it would be partly legalized—with passage of certain federal acts. One was a railroad act meant to foster expansion of the nation's rail system. If you said you planned to build a railroad and went through certain motions of preparing to do so, a grateful federal government would give you the right-of-way of your choice over public land. This consisted of twenty square-mile sections of public land for every mile of track laid—or for what appeared to be preparations for laying track. The sections of land granted for these rights-of-way alternated with sections of public land that were usually assigned to local school districts.

Many of these "railroads"—most of which had the magic word "Pacific" somewhere in their company name—never contributed anything to the nation's rail system. Mostly, they contributed log traffic to the rivers, for once a foothold was gained in a right-of-way, the lid was off. When those sections owned by the railroad company had been shorn of their timber, the lumberjacks simply moved onto the adjacent sections of public land. Each section of timberland acquired more or less legally by a railroad company meant that at least one, and possibly six or eight, sections of public land would also be cut. As local school districts came into being ten or twenty years later, they would find their "school lands" stripped of the best timber.

Not that you had to go through the motions of building a railroad to do all this; a little later, you could get the job done through the Stone and Timber Act.

This was another of those bursts of federal generosity such as

the one that kindled the Homestead Act, and, like the latter, was
intended to foster settlement of public domain "wilderness." Under
this act, it was possible for an individual to enter a claim for forty
acres of government-owned timber. To block any timber-grabbing
by corporations, it also provided that eight persons could file on
160 acres of timber—provided that none took more than twenty
acres for himself. This was reasonable enough and might even have
cleaved unto the public weal had there not been a yawning loop-
hole: the claims could be filed by power of attorney. So, the letter
of the law was duly observed and powers of attorney were filed—
and timber claims were entered in the names of lumberjacks who
probably knew nothing about it, dead men who certainly didn't,
and butlers, camp cooks, dogs, horses, and cats. The daughter of
a certain lumber baron had a pet monkey that became the owner
of a great deal of fine timberland under the name Simian Wynwood.

As on the railroad sections, once a forty-acre timber claim was
established it was usually open season on adjacent tracts as well.
The original, quasi-legal claim was a way of getting your foot in
the door—and your men and their axes into someone else's woods.
Not only were public lands trespassed by cutters, but many private
claims were jumped as well. For awhile, the only practical way to

deal with such claim-jumping was with gun and axe handle. Taking the matter to court was a waste of everyone's time as a rule, because a timber thief was not likely to be convicted by local jurymen who were either stealing timber or planning to.

At the peak of the cutting in the 1880s there were about 140,000 lumberjacks in the winter forests of Minnesota, Wisconsin, and Michigan. It was dangerous, cold, wet, brutalizing drudgery that began each workday in the dark and ended in the dark. They sought out the finest groves of white pines; early in the logging game, they were taking no logs that were not eighteen inches in diameter at the small end—and four sixteen-foot logs were often gotten from a single tree. The logs were stacked in great decks on the banks of frozen rivers, or even on the edges of the ice, and in spring the river pigs would cant the logs into the river and begin the downstream drive to such places as Beef Slough—a branch of the Chippewa that entered the Mississippi just above Alma, Wisconsin—where the logs would be scaled, sorted by ownership, and made into rafts. It was a classic Upper River industry, for those logs rarely went any farther south than Saint Louis.

Rafts were built of logs or rough-sawn lumber. The log rafts were relatively simple, composed of sections or "strings" of logs bound together with heavy pegs and lashings of birch and burr oak withes. A lumber raft, on the other hand, was often an elaborate masterpiece of backwoods engineering. The basic unit of such a raft was the "crib," made of one-inch planks pegged and bound into a sixteen-foot square that was twenty-four planks deep. Each crib could hold about six thousand board feet of lumber and was joined to others to form a "string." Some of these lumber rafts contained enough pine boards to build a small town; in 1901 Captain George Winons and the steamboat *Saturn* brought a lumber raft from Stillwater, Minnesota, to Saint Louis that was about one city block wide and five football fields long, and held nine million board feet of rough-sawn white pine lumber. Because log rafts were usually only one course of logs deep, they represented far fewer board feet of lumber. One of the biggest on record was towed down the River from Lynxville, Wisconsin, to Rock Island in 1906 by two steamboats—almost ten acres of logs in a raft 1,550 feet long containing over two million board feet of lumber.

Rafts might be brought downriver by the same men who cut the trees, or they were handled by specialists. In any case, raftsmen were a breed apart, and as fine a bunch of buckos as ever gouged an eye or wrecked a saloon. Reincarnations of Mike Fink to a man, they were universally regarded as ramstugenous alligators who played about the same role in the river towns that the wildest of the Texas drovers filled at Kansas railheads.

Charles Edward Russell, later a riverman in his own right, grew up in Davenport, Iowa, during the heyday of rafting and knew whereof he spoke:

> A raft made fast to our peaceful shores would stir sober citizens to uneasiness and fill us children with vague alarms. Experience had somewhat overtutored the world of the law-abiding in the ingenious deviltries of raft crews; yet there were gradations in quiet. Raftsmen from the Chippewa, the Wisconsin, from any other stream might be violent, crapulous, reckless, and boisterous; by common consent they were not without a human substratum. Only raftsmen from the Black River [Wisconsin] were known to be always the true sons of perdition of the rip-roaring branch of the family tree. With them, according to common belief, rafting was but a diversion; the real business of their souls was battle, murder, and sudden death.

This was common knowledge, said Russell, based on such unimpeachable authority as Deacon Conduit, who kept the hotel at Le Claire and had been to Saint Louis three times. "A raftsman would just as soon stab you as look at you," opined the Deacon. This opinion had wide currency among such good folk as the Widow Fowler, who always hid her silver spoons when a raft was tied up at Guttenberg Bend.

When the rafts were delivered and the raftsmen paid off, they were turned loose. But rough and hard as they were, there were even harder predators lying in wait for them in riverfront dives and on the floating deadfalls called "love boats" that were usually anchored not far from delivery points. These were the cultural forebears of television's "Love Boat," with some of the same philosophical tenets but none of the finish. A Mississippi River love boat might be a couple of one-story buildings on a pair of joined flatboats, specializing in the rawer kinds of Monongahela whiskey, dancing, gambling, and harlots. They were basically brothels with added attractions, peopled with thieves, bully boys, crooked gamblers, and harpies of the worst sort.

The Upper Mississippi, with its dark backwaters and twisting byways, was a wild place then, an interface between civilization and the fading wilderness—a frontier existing through the end of the nineteenth century in an otherwise tame and civilized region. Most of that wild River, of course, was interstate water. It was an easy matter for a love boat to shift operations a few hundred yards and be out of a state's jurisdiction. This was attractive not only from a business standpoint but from a personal one as well; many of the tenants of those river hells were on the dodge. For that matter, so were some of their customers; there were a substantial number of nineteenth-century lumberjacks and raftsmen whose presences were urgently required elsewhere, and there were few better places to slip the bonds of law than remote lumber camps or on rafting crews. But rough and tumble as these men were, they were nonetheless amateurs, while those running the love boats were as professional, efficient, and unprincipled as sharks. "About the only good thing to be said for the predaceous hosts and hostesses," wrote Russell, "is that they used to sing 'Buffalo Gals,' which is a good song."

The early raftsman and his lumberjack connection were civilization's expendables. Such men were always out front, far ahead of church and law, probing and testing the wilderness and sending back its gold, fur, and lumber to build civilizations in which the men themselves were never really welcome. Soon they would no longer be needed—and would follow the wilderness to wherever they had helped drive it.

Yet, even as these spoilers destroyed the northern pineries and scandalized good folk from the headwaters of the timber rivers to Saint Louis, they helped create the cities, towns, and farms of the heartland with their lumber. On the central Iowa farm where I spent much of my boyhood there is still an old barn that was built in the 1870s. The main beams spanning the eaves of the hayloft are rough-sawn white pine, twelve inches square and a clean, knot-less thirty feet long, pegged, not nailed, to the uprights. I once asked my Uncle Cliff Ross about the timbers in his barn, and all he could tell me was that they had been "brought from the eastern part of the state." I've no doubt that they were floated down the River to Davenport or Clinton and then freighted inland to frame a proud new barn in the decade after the Civil War. That barn would see thousands of milkings by lantern light, with generations of cats patiently awaiting the stream of milk warm from the teat. It would hold countless tons of fragrant hay, hundreds of suckling calves and lambs and laying hens, great gentle draught horses at rest, children leaping from the loft beams into the deep hay, and little girls searching for new kittens hidden in the secret places by the old tabbies. It would bear a century of work and love, hope and laughter, all grown out of a few cribs of lumber brought out of the north by some half-savage, vagrant men.

As successive components of a swelling commerce, the flatboats, keelboats, steamboats, and timber rafts had ruled the Upper Mississippi from the 1820s until the Civil War. The late part of this period—the 1840s and 1850s—was the Golden Age of the paddle wheelers. But the last third of the century would belong to the steam railroad locomotive. Although steamboats had a flush of activity after the Civil War, and looked for a time as if their commerce would be as vigorous as ever, their day was passing. The Golden Age of rail was at hand.

Even as steamboating faded on the River, timber-rafting continued—although in its latter day the huge rafts did not float free but were "towed" (pushed, actually) by steamboats. As a viable element of river commerce, timber-rafting survived a little longer than the steamboat industry. But the vast forests of the north had been wantonly spent, and by the early 1900s the rafting years, too, were almost over. The last lumber raft to be formed on the Upper Mississippi was put together in the summer of 1915 from remnants gathered at closed mills. The raft was built at Hudson, Wisconsin, on the Saint Croix River, towed by the *Ottumwa Belle*, and delivered at Fort Madison in August. Then it was all over. An old raftsman later calculated that rafting on the Upper Mississippi had

brought 46,970,000 board feet of white pine to the downriver mills and yards. The little remaining lumber from the forests of the Upper Midwest was now being shipped by car, not river, and the bulk of the nation's lumber began to come out of western forests.

With paddle wheelers succeeded by railroads, and the northern pineries exhausted, the Upper Mississippi quieted. River landings vanished, casualties of floods and neglect, and were never replaced. River towns that had been born of steamboat commerce dwindled and slowed and fell asleep to dream of the past. Old river captains sat on their porches and watched downstream for steamboats that would never return, singing the past glories of such lightning boats as the *Key City* and *Grey Eagle* and the tobacco-juice heroes who crewed them, and mourning the lost loveliness of such as the *Lady Franklin*.

To the steamboatmen, merchants, and townsmen whose fortunes were mortally stricken by the fading of the main chance, life on the Mississippi had ended. But there were others, spiritual heirs of the red men who had built those effigy mounds on the headlands, for whom life on the Mississippi was just beginning. They built their johnboats, invented a strange device called a crowfoot bar, and went treasure hunting.

2

SHELL GAME

1

Joe and I were in Katie Quillan's café eating pork chops and I had offered to pop for supper, which I didn't mind doing but was just as happy that it didn't happen every night. For no bigger than he was, Joe Martelle was a great feeder. He never encumbered himself with much breakfast or dinner, didn't dull his appetite with nicotine or take on any empty calories with alcohol. He saved up his hungries for supper—the last meal of the day in Harper's Ferry, Iowa—especially if he knew I was springing for it.

1   2   3 4   5

He had been eating with both hands for a steady half hour, and even before he had finished his first wedge of Katie's exalted apple pie he had signaled for another. She had been expecting that, and had the pie waiting. Joe dispatched it with undiminished fervor, pushed the plate away, reached for his third cup of coffee, and beamed across the table at me.

He had washed up before supper and put on a clean, faded, hickory shirt, but he still wore two days' growth of white stubble and the inevitable pair of hip boots. With those boots and two pieces of pie, Joe may have gone a hundred sixty pounds and even then, well into his fifties, he was still built along the lines of a good varsity pole-vaulter. He moved and handled himself like a man

1. washboard
2. pig-toe
3. Higgin's eye
4. spectacle case
5. three-ridge
6. three-ridge
7. pocketbook
8. squaw foot
9. butterfly
10. buckhorn
11. wartyback
12. elk toe
13. pink heel splitter
14. pink paper shell

6   7   9   11   13   14
        8   10   12

half his age, but any illusion of youth ended at his face with its rough, deep-grained brown so unlike any mere beach tan, and at his hands—calloused and broken-nailed with fingers thickened from a lifetime of heavy uses. We never shook hands when I was not conscious of my city-soft paws and a bit ashamed for them, something Joe could never have comprehended, for he respected a man who worked more with his head than with arms and back.

Joe wasn't the smartest guy in the world and didn't exactly burn with conventional ambition, but he was smart where it counted—out there on the River between Whiskey Rock and the mouth of Big Paint Creek. He batched it in a little cabin on the bank of Harper's Slough in extreme northeastern Iowa and as far as I knew he had never held a regular job, punched a time card, or worn any man's collar. He was a riverman pure and simple—trapper, commercial fisherman, crowfoot-bar sheller of the old kind, and when he wasn't on the River he would likely be up in the big rough hills behind town cutting white oak staves for whiskey barrels or hunting timber rattlesnakes for the fifty-cent bounty.

There was something feral about him. No, that's not right. Feral infers tameness gone wild, and Joe had never been tame or gone really wild. There was just something, well, *riverish* about him, and that figured, since the River was his tutor, provider, and main entertainment. His concentration had never been diluted by such trivia as reading, television, automobile, or wife. I doubt if he had a social security number and I'll lay odds that he never paid any taxes. From any bureaucratic point of view, he scarcely existed. The net result of all this was a sublime social innocence, and a profound sophistication in the ways of the River.

Joe was a gentle man, or so he always impressed me. I never saw him angry or even very impatient and that fit the pattern too, for the river world demands forbearance and endurance, and the lack of either quality will at least cause a man to fail and may even put him under. There may have been another side to Joe, however, a darker vein seen only by those who had known him all their lives.

Jim Williams was the local tavern owner and fishing-tackle dealer, and since just about every man in town drank and/or fished, and usually talked a lot in either case, Jim knew where most of the local bones were buried. He and I were fishing the points of islands in McDonald Slough one day, casting for bass in the brush and stick-ups, when Jim spoke of Joe's father.

"They found him laying in his drifting johnboat, back on one of the sloughs. He'd been all shot up. Happened years ago."

"What was that all about?"

"Well, there was talk that he might have been tending someone else's traps and trotlines. Stuff like that. Whatever it was, somebody took it real personal."

"Damned if somebody didn't. How did it all come out?"

"There wasn't any arrest, if that's what you mean. So there was no conviction, neither."

"No clues? No leads or anything?"

"Well, there are local folks who might know things that never came out. Maybe the killer still lives around here—with nobody ever talking about it. I say *maybe*, understand. And maybe that's because folks figure that Joe would go after his dad's killer if he knew who it was, and maybe they think too much of Joe to see him cutting or shooting somebody. A lot of maybes. Anyway, that's where it's laid all these years. That's where it's always gonna lay."

But there in Katie's café, full of pork chops and pie, Joe radiated good will. He beamed upon the world in general and me in particular. "Well," he said. "That was all right. Much obliged. I was gettin' a little high in the flank."

"What's doing tomorrow?" I said, trying to keep his mind off another piece of pie. He still had a peckish look.

"Oughta be good on the river," he answered. "It's on the rise. You want to go out and pick some shell?"

"You bet."

"See you down at my place, then. Be there in good season. I want to be done cooking out shell by late afternoon."

"What time you want me to be there? Six-thirty or so?"

That wasn't the brightest thing I've ever said, and Joe made the most of it. He enjoyed such questions from me, and I kept him well supplied.

"Six-thirty? Did you say *six-thirty?* Farmer's hours! You want to lay in bed like some old farmer or do you want to go out on the river? Be down at my landing at four-thirty."

He finished his coffee and slid out of the booth and stood grinning down at me in mock despair. Again, as so often before, I was struck by his well-used quality. The faded shirt turned up to the elbows of corded, sun-blackened arms, the jeans stuffed into folded-down hip boots, also patched and hard-worn, their leg straps cut off so that the boots could be kicked off in deep water. But it was mostly the face, that lined and weather-stained face with the strangely pale forehead where his hat came. Why, Joe's getting old, I thought. Old and poor. And then a quality came through as it always did, the sense of a country boy about to go skylarking and scarcely able to contain it. Joe's tomorrow look, his going-out-on-the-river-with-John look, knowing there would be talk and jokes and just having someone along for a change. And I thought again: Who's old? Who's poor?

I told him that if he'd let up I might bring some sandwiches, like Katie's roast beef kind, on homemade bread. Just in case we got a little hollow on toward noon.

"And a couple pieces of pie?" he added. "Katie could do it up in wax paper or something. We'll need it to celebrate. We're gonna find that big pearl." He turned to leave, with one last despairing shake of the head.

"Six-thirty. Laying in bed half the day . . ."

A high-pressure cell had cleared the night sky and chilled the air, and in the gray pre-dawn the surface of Harper's Slough was overlain by a blanket of mist that looked thick enough to walk on but was beginning to lift and thin as the light strengthened. Over on Delphy's Island it was still night back under the river birches and huge old silver maples with their arm-thick vines of poison ivy, but here on the west bank it was lighter and I could see Joe down ahead working hip-deep in stirred mist, laying crowfoot bars on the racks of his johnboat. None of the smells had yet been burned away by sun; they had been sharpened and intensified by night coolness and dewfall—the dawn smell of dampened road dust, the work-smells of wood ashes under the shell-cooking tanks and fish smokers along the riverbank, and gasoline and tarred nets, the sweetish reek of clam meats and a large, very dead fish somewhere nearby, and underlying it all the rich organic smell of the River itself, compounded of mud and water, rotting wood and leaves, wet sand, mink urine, and the whole infinite roster of dead and living components that constitute the wild Mississippi. Birds were awake now and beginning to sound off; not far away a prothonotary warbler was celebrating a hollow nest snag. The crows were up and about their raffish doings—flying the black flag of themselves, as the poet said—and out on Delphy a barred owl was making his last statement of the night. For the hundredth time I thought this must be the finest part of a summer day, when everything is fresh and unused and not torn by human affairs and the people who are up seem to go about their work quietly and with respect, as if to make that time last as long as possible. Like Joe down there on the riverbank, grinning his welcome and saying something about pie as I walked down the path to his landing.

There wasn't much about his outfit that was "boughten." He had built the boat himself, probably saying as he did so that "poor folks have poor ways," making a classic river johnboat about eighteen feet long with flat bottom and squared bow, with a short work deck up front. As I recall, it was largely cedar and white pine planking with white oak gunwales and transom and powered with the classic ten-horsepower Johnson outboard motor—then the standard for such medium commercial fishing craft.

On each gunwale was mounted a set of notched wooden uprights on which rested the crowfoot bars, or "brails," birch poles festooned with short lengths of chain. From each short chain hung one or two of the "crows' feet," small blunt grapnel hooks fashioned of heavy wire. As the boat drifts with the current, one of these crowfoot bars and its ranks of hooks is dragged along the river bottom and over beds of freshwater mussels.

In the Upper Mississippi, half-buried in silt and sand, are scattered congregations of naiad mussels. They are simple creatures, little more than two strong shells or "valves" enclosing the soft, formless body. Blind and virtually brainless, they lie on the river bottom with shells agape, laved in the currents that bring them food and oxygen. These feeding mussels are sensitive to disturbance and will clamp tightly shut if touched by a foreign object. When a crow's foot hook is dragged over one of the gaping mussels the irritated bivalve seizes the hook. In due course, as the sheller feels he has floated far enough over a mussel bed and the bar begins to feel heavy, the crowfoot bar and its clam-encrusted hooks are heaved up onto the boat racks and picked. The sheller always carries two crowfoot bars and sometimes three, with one always in the river and working. Joe's bars were hand-trimmed birch poles cut from a rocky hillside above town; he scorned poles of metal or heavy kiln-dried wood.

"They got no spring!" he would declare. "My kind of poles are the best on the river. Know why? They got some spring to 'em, and the river shapes 'em to the bottom so's the hooks get down and work in all the holes and places. Most bars are too stiff and dead, with no give."

We came to the end of Big Swifty Slough and Joe turned into the current and cut power; for a moment the boat was dead in the water, then the current began to take hold and Joe said judiciously: "Like I said last night, there's a good current today. We'll start here. Put out the mule." One of my several jobs was tending the "mule," a crude kite of wooden framing and light sheet metal that rode in the water from a harness attached to the bow of the boat. With minute adjustments of the harness, the mule deflects the current at various angles and the drifting boat can be shifted back and forth cross-current. A cunning old river trick—steerage without power.

I set the mule to Joe's satisfaction while he checked the larboard bar's rigging, took off his old denim jacket, lifted the crowfoot bar from its rack and dropped it into twenty feet of brown water. I wondered about that. We were working on the shadow side of an island where there was still some mist and the pre-dawn chill, and I asked him why he'd removed his coat.

"Look at them hooks and all that heavy chain. What if I was out here alone and snagged my coat when I dropped that bar into the river?"

Joe had been on the Mississippi a long time and hadn't drowned so far. I looked at the ranks of shiny, sand-scoured hooks and took off my jacket.

With the first bar down and working, we relaxed for a hundred yards as we drifted through thinning river mist. Then it was time for the first lift of the day, and the work began. Sometimes, when barring a very rich bed, there may be one or two mussels gripping every hook and sometimes even the chain. This could add up to a hundred or more big clams even on a fairly small outfit like Joe's, meaning that he might swing his own weight of crowfoot bar and clinging mussels up onto the rack.

Like most solitary fishermen and trappers, Joe was either dammed up or in full flood. No one can outtalk a habitual loner when he's loosened up by congenial company, and as we worked down the bed of seven-inch "washboards" he grew happier and more loquatious as the bars ran heavy, his talk flowing bankful.

"Hey, look at that shell! Big shell that runs heavy. Don't mess with them little pocketbooks and pig-toes. The big stuff is what we want. You pick and I'll get this other bar to working. Lookit them big old devils, would you! That's old shell. Say, we'll sure check the meats for pearls when we cook out. Watch out for that rope. Didn't I tell you the current would be just right? They had a helluva rain up by La Crosse. Lemme get past you and fix that mule. Hey, did you see this one here on the end? They don't come any bigger than this! You ever see the like?" All with a broad grin, the wrinkles fanning over his face.

On such days—and they were rare enough even if you knew him well—he would keep up an unending flow of talk about the river and its sloughs, clams and their pearls, birds and critters along the banks, and storms on the great river pools. He might talk about the old river jungles of Camanche and Sabula; of the knifings, fur thieves, and drownings—accidental and otherwise. He would tell of bounty-hunting rattlers in the limestone outcrops of the hills, and of the secret caves he'd found there. He might speak of all the things on the River that surprise and bemuse and instruct a man, and as he talked he swung the heavy crowfoot bars up onto their racks and filled the boat with money shell.

Then we'd be at the end of the bed or what Joe judged to be the end, and he would direct me to stable the mule while he picked the final crowfoot bar before yanking the starter cord and heading back upstream to begin another drift. We had started at sunrise and were still at it at noon—seven hours of almost steady labor, mostly for Joe. It didn't seem to bother him. And although my only job had been handling the mule and helping pick shell, my hands were already sore from the tarred line and from twisting the mussels off the hooks.

We tied up to a soft maple that leaned out over the slough at the head of the run and ate our beef sandwiches and apple pie— soggy, badly abused apple pie that had a slight flavor of tar and river mud.

"Over there a ways is a little bar you can't see," Joe said. "I got my first shell there when I was a kid. Pollywogged for it. Waded around barefooted and felt for the clams and ducked down and got 'em and stuck 'em in a gunny sack. There was a lot of shell back then before they fenced the river."

"Fenced. I never heard it put that way."

"Well that's what it was when the big dams went in. A lot of the old clam beds was lost. Changed the river so's you wouldn't know parts of it."

"You ever trap this stretch?"

"Not right through here. Mostly upstream. Did pretty good on mink last winter but prices weren't worth a damn. Hey, I ever tell you about racing a otter a couple of winters ago?"

"Come on . . ."

"Yeah, I did. It was in wintertime. If there's good ice maybe I'll run the whole trap line on skates. Nobody else does and you can sure get to where you're going. There'd been a cold snap with no snow and the ice was thick and real clear and I saw something moving under the ice just ahead of me and there was this big old otter! I put on some speed and so did he and I stayed just behind him. He was every bit of four foot long, that old otter. Damn, can they go! No wonder they can ketch fish. I stayed with him for oh, I don't know, maybe from here to that piss ellum over on the bank and then he turned off and I lost him. You ever heard of anybody else doing that?"

I had not.

"Well, I sure did. Raced a big old boar otter. You ought to

write that up sometime. I bet nobody else on this river has ever done that."

He washed his hands in the slough, splashed some water on his face, then cupped his hands and drank several noisy draughts. "Ain't you gettin' dry?"

"Not that dry."

"Suit yourself. Better than you get in town. Most town water tastes like medicine. This is real water. A little mud never hurt nobody."

The boat was riding lower now, with no more than ten inches of freeboard between the river and gunwales. And still Joe wasn't ready to hang it up. "It ain't every day I got free help," he explained, "and we got time to make a couple more drifts." We made only one more as I recall and then, with almost eight hours of steady work behind us and the gun'ls nearly awash and a small hill of washboards, three-ridges, and muckets between us, Joe pointed the heavy boat homeward. Cookout time.

Just below his cabin on the bank of Harper's Slough was a wood-and-sheet-metal tank with a firepit beneath. Joe quickly built a fire of driftwood, added a bucket of water to the tank, and as it began to steam we poured in a couple of bushels of mussels and covered them with an old piece of canvas to seal in the steam. The object was not to thoroughly cook the mussels but to simply kill and relax them until the shells opened and the meats could be easily scooped out.

We soon drew some company—a portly black Labrador retriever that took up station a few yards away and awaited any forthcoming charity. The steam leaking through the canvas began to carry the heavy, cloying smell of cooking shellfish that I've always found oddly unpleasant. Then the canvas was thrown back, revealing the gaping mussels. Joe began to steadily sort through them, pitching the shell into various piles with one hand as he scooped out the meats with the other, dropping them into a large bucket after he had carefully kneaded them to see if any contained pearls. Some of the flaccid meats were pitched to the old dog, who fielded them expertly until he'd had enough, and then walked heavily up the path to a pool of deep shade where he collapsed with a sigh and immediately fell asleep.

"Joe—you suppose that Lab knows something we don't?"

"You ever try to eat any clam meats?"

"No. I've eaten about everything else in and around a river, but never clams. How about you?"

"Tasted 'em, is about all. Pork chops are better. They say Injuns used to eat 'em in the olden times. Maybe so. I've never known any river folks to eat 'em and I've known some hungry river folks." His swift hands never paused, scooping out clam meats and checking them and then grading the shell according to weight and kind, a job I could not help him with.

For the best thick, commercial shell Joe was getting something like forty dollars a ton, back there in 1953. For thinner shell, much

less. On an exceptional day he might take a half-ton of shell of all kinds, a catch for which he would be paid fifteen dollars, tops. His day's catch was sacked and labeled, and left trustingly for the buyer who came around every couple of weeks with a truck. No one ever stole anyone's sacked shell around there, that I ever heard. To do so would be a shocking infraction of several codes, with a sliding scale of retributions that were gauged on the reputation of the malefactor and the temperament of the offended party.

As nearly as I could ever figure it, Joe's daily labor with crowfoot bar and cookout tank produced an average hourly wage of something less than a dollar and a half. Of course, that was in the early 1950s and Joe's gross and net incomes were also a lot closer together

than most people's. But even with his low capital investment and negligible overhead, there were risks and losses that hurt. Of all the shell caught even on a good day, much would be so thin and worthless that it was discarded. Some of the valuable varieties would be too small to keep, and were returned to the river. There would be downtime with damaged equipment, such as a crowfoot bar fouled on some ancient sunken log. The snag might be raised bodily if it were not large or fast in the river's bed; if not, several chains might be torn away, or the entire crowfoot bar lost. Through the years Joe had lost several crowfoot bars to heavy snags and he hated that, but even more galling was the knowledge that some of the best remaining clam beds were in old sloughs choked with sunken stumps and trees.

Long before any scuba was in use on the Upper River, Joe could be turned on with talk of a home-made diving outfit.

"There ain't all that much to it," he said one night in Jim's tavern. "Look. All we'd need would be some sort of air pump and a helmet made out of sheet metal with some glass in the front of it. It wouldn't be hard to fix the helmet so's it wouldn't leak. Why, we could strip them old sloughs. I know places that ain't been touched in fifty years!"

"Yeah. I could handle the air pump. You know more about shell than I do."

Jim Williams was stacking beer mugs and had his back turned to us, but I could see his shoulders beginning to shake.

"Well," Joe replied, "I ain't as young as I was and I sort of figured I'd handle the pump and you'd be in the water."

"No."

That's as far as it went, although Jim stretched it out into a pretty good story as time went by. But it didn't take much imagination to conjure up a fairly bad vision of what might wait down in the dark, tangled depths of a wild Mississippi slough. I could still remember when I had dived to free the lead line of a bag seine and gotten caught in the roll of discarded fence wire that had snagged the net. It was a thrill that lasted for only twenty seconds or so, and that was long enough.

A man needs some dreams, though, and one of Joe's was of a snag-filled slough with a virgin clam bed that held the Pearl—that breathtaking unbelievable one pearl he had hoped for most of his life. For in the back of every sheller's mind, however pragmatic and stolid he might be, there is always that thought of the great pearl.

No one knows if the river Indians hunted pearls or just found them in the course of gathering mussels for food and as tempering grog for their fired earthenware. But there can be no doubt that pearls caught their fancy and were valued. Burial mounds in Ohio have contained thousands of the chalky, crumbling remains of bored pearls that had been necklaces of rank or adornment. Some Indians inletted freshwater pearls into the canine teeth of bears that they so prized as trophies and ornaments, and when the early French explorers came to the Upper Mississippi they met Indians who wore ear pendants and necklaces of fine pearls.

Surprisingly, there doesn't appear to have been any pearl fever of record among white men until 1857 when a shoemaker named Howell happened to gather some mussels for his supper from Notch Brook near Paterson, New Jersey. That he was dining on mussels is somewhat notable; that he bit into an excellent four-hundred-grain pearl during supper is downright singular. The pearl's gem quality suffered from the boiling and biting, and a jeweler told the cobbler that it might otherwise have been worth thousands of dollars. In a time when a working man could support a sizable family on a few hundred dollars a year, this was heady stuff. People descended *en masse* on poor little Notch Creek and within two years there wasn't a mussel left—but the local citizenry had found about $115,000 worth of gem pearls.

Such news travels fast, and people took a new look at their local streams and rivers. Pearls were soon being found in Pennsylvania, Ohio, Texas, and Arkansas, but there were no spectacular finds until some boys playing in the Little Miami River in Ohio came home one day with a handful of pearls of delicate rose pink. The little stream yielded $25,000 in pearls. A few small finds occurred in Florida, Vermont, Kentucky, Tennessee, and as far as Washington, but pearl fever cooled until 1889 when the Pecatonia River in southwestern Wisconsin began to produce pearls—about $10,000 worth. Pearls were soon being found in other Wisconsin rivers:

the Sugar, Apple, Wisconsin, and then the Mississippi near Prairie du Chien. It was shelling unrefined since the time of the Mound Builders: the pearl hunter simply waded barefoot into a stream, searching by eye in the shallows and feeling with his feet in deeper water for mussels, which were cut open, quickly inspected for any pearls, and then discarded. Crude and inefficient, maybe, but they were working primeval beds that held large stocks of the old mussels that could be expected to produce the finest gem pearls. By 1891 about $300,000 in pearls had been found in the Upper Mississippi valley alone—and the Great Pearl Rush reached its zenith in 1902 with two pearls found near Lansing, Iowa, that were reportedly sold for $50,000 and $65,000. The more valuable of the two was said to be almost an inch in diameter and of flawless color and form.

When I first met John Peacock, in the early 1950s, he was a small, balding, elderly gentleman lately retired from Prairie du Chien banking. As a young man during the heyday of the Upper Mississippi's mussel shell industry he had been a pearl buyer— one of that sharp-eyed band of speculators and entrepreneurs who, like most rivermen of the day, were drawn by the lure of the great pearl. One of the few old-time Mississippi pearl buyers then alive, John Peacock told me that at one time there were twenty-seven pearl buyers registered at the Prairie du Chien hotel—agents from India, France, England, and all parts of the United States.

Eighty years ago, when the Upper River boomed with mussel fishermen, such buyers would have breakneck horse races along rough bottomland roads when word reached town that some clammer had found a fine pearl while cooking out shell. Although the fisherman might be stunned by the magnitude of the first offer, he usually managed to play the dickering game. His wife, on the other hand, might be completely undone. For the first time in her life she was within reach of a new stove, store-bought dresses, all manner of heady wonders seen in the mail-order wishbook—and, headiest of all, if a particularly fine pearl was involved, even a new frame house with lace curtains. Peacock remembered going to a dirt-floored, three-room cabin where a rather good pearl glowed from a scrap of flannel on the kitchen table. He offered five hundred dollars and was prepared to go higher, but the fisherman's wife couldn't take the pressure of the slow and cautious bargaining. She dragged her husband outside and issued some manner of brief, emphatic directive—and the fisherman hurried back to accept Peacock's second offer.

One night in 1907 a Harper's Ferry bartender sent an urgent message across the river to Peacock, who had a reputation for fair dealing. A fisherman was in the saloon sopping up cheap whiskey and offering a fine pearl for sale. After a day and night of drinking and dealing, his price had dropped with the local whiskey supply until he was finally asking only a few hundred dollars. None of the locals could afford the pearl even at that price, and the man's friends feared he would lose the pearl to a thug or some slick-dealing outlander.

Peacock arrived the next evening, getting there well ahead of any other buyers, and while a dozen fishermen looked on and the whiskey talk died down, he screwed his jeweler's loupe into his eye and made his first appraisal of the pearl by lamplight. Next day, when the fisherman was sober, Peacock paid him a thousand dollars—which, in that time and place, was enough to sweeten any hangover.

"I look back and see that as a risky thing," Peacock told us one day as we sat on the little dock near Joe's cabin. "Mighty risky. Appraising that way by the light of a coal-oil lamp, I mean. Oh, it wasn't the biggest pearl that had come my way, but it was one of the finest. Pearls are rather like women, though; they often look their best by lamplight. When I saw that pearl by full daylight the next morning, though, I knew I'd been right. The pearl went to a Chicago dealer for five thousand dollars and, as I recall, it eventually ended up in the New York market."

Some finds were the blindest kind of luck. A fisherman near Prairie du Chien had bought some clam meats, which can be good catfish bait at some times of the year, for use on his trotlines. He came home in the afternoon with his catch, cleaned the catfish, and threw the offal to the chickens. As he sat smoking his pipe on the back steps he noticed a scrawny rooster having trouble with some of the catfish innards. The man went to the rooster's aid and found the problem to be a section of catfish intestine that contained a large obstruction that the bird was unable to swallow. It was a large, rough, discolored pearl, and when the man saw Peacock a few days later he leaped at the buyer's offer of $150.

Pearls are composed of concentric layers somewhat like onions, and bad surface blemishes can sometimes be peeled away with delicate instruments and a rock-steady hand. Peacock had done this often enough, and frequently with gratifying results, but he had little hope this time. The pearl was rough, dull, blotched, and discolored; the best that could be said for it was its size and excellent symmetry. Other than that, it was just an ugly lump of calcium and assorted minerals.

Late one night, when his house was silent and he could give full concentration to the delicate task, Peacock carefully peeled away the blemished outer layer to reveal the rarest gem of his career—a pearl of rich pigeon-blood red.

When the old gentleman had finished this account there was the introspective hush that always attends a telling of The Big Find, whether it's a vein of gold, an egg-sized diamond, or a gorgeous nymph who owns a liquor store. Joe was looking at Peacock with the hopeful reverence usually worn by supplicants at Lourdes, while Jim Williams was gazing off down Harper's Slough in the manner of an old fox remembering his last fat pullet and dreaming of pullets to come. I was still awaiting the bottom line, which didn't seem to be coming.

"May we then conclude," I finally asked, breaking the general reverie, "that you managed to recoup your investment in that pearl?"

"There was no difficulty in finding willing buyers," Mr. Peacock replied with a dreamy smile. "I did manage to recoup my investment. Oh, yes. I did indeed."

That's as much as we ever found out, but something in his tone led me to suspect that he had entered the banking business the day after he'd sold the red pearl.

Color tones of freshwater pearls are dictated by the mother shell, of course. The big washboard mussels usually have pink pearls, as do the wartybacks. The three-ridges have pearls in shades of blue, green, and lavender; from muckets come fine pink pearls, and the sand-shells have pearls of salmon-pink. But from the delicate little ladyfingers may come the most beautiful of all—the prized black pearls with slaty fires of gray and violet iridescence.

Wild pearls occur in shapes of balls, buttons, pears, petals, and leaves. The classic sphere is produced when the developing pearl is located where it is free from unequal pressures and much movement. A flatter button pearl is usually formed near the outer surface of the mantle—the membranous organ that covers the inner surfaces of both shells and secretes the nacre that comprises mother-of-pearl—where the developing pearl is pressed against the shell. A misshapen pearl or "baroque" may result from a parasitic infection of some sort, or from a button or spherical pearl that has been shifted from its original location. Such baroques may be free in the mussel, or attached to the shell—which is the only way I've ever found them. Most of the pearly concretions found in mussels are baroques, a point of considerable personal pleasure since I've always preferred the free sculpture of a fine baroque to the symmetry of a spherical pearl. I'm apparently not alone in this. The other day I received the catalogue of Boring and Company, San Francisco, who specialize in fine Asiatic freshwater baroques. As the copy points out: "The perfect pearl need not be round. The perfect pearl need not be white. The perfect pearl need not be traditional." Over the years I've known of lovely baroque pearls that were casually given away by the finders, but there's nothing casual about Boring and Company. One model of their single-strand necklaces lists for $9,800.

Joe had a handful of baroques that had some market value, although their main use at the time was for costume jewelry. He also owned several small pearls of good shape and color, and one morning Peacock affixed his old jeweler's loupe and appraised them.

"Very nice," he said. "Good color, texture, shape. Pity you couldn't have shown them to me fifty years ago."

The gem pearl resource of the Upper Mississippi was a casualty of technology. The best and largest sweetwater pearls are usually produced by older, larger mussels. Thick-shelled species of river mussels may commonly live from twenty to forty years, and almost a century in some cases. Compared to pristine numbers, however, only a relatively tiny proportion of the older-year classes of mussels have managed to survive intense overfishing, changes in the river-bed wrought by wing dams and channel dams, and general pollution and siltation. As these old-year classes dwindle within the general mussel population, chances of finding large, fine pearls dwindle apace.

Mississippi pearling has also been a victim of Oriental ingenuity. By creating cultured pearls in oyster beds under carefully controlled conditions, the Japanese were able to supply good matched pearls and sharply undercut the value of the rarer wild pearls. A pearl's price, after all, is a reflection of the number of man-hours invested in it. Infinitely more time and effort are required to find a perfectly spherical wild pearl than to culture its tame equivalent, and a necklace of forty-grain pearls that might be bought today in a discount store would have cost thousands of dollars at the turn of the century. A twenty-grain pearl found near Harper's Ferry when Joe was "barring the beds" there brought only $120—a fourth of its value in 1910. At the time John Peacock appraised Joe's pearls, they would probably have averaged no more than $30.

Some pearl fanciers have said that rivers lack many salts and minerals found in the ocean and that river gems are softer and duller than marine pearls. Peacock agreed that cultured pearls were of high quality despite their relatively low price. But he would brook no mean-mouthing of his beloved Mississippi pearls and contended they would equal any produced in a Pacific lagoon, usually considered premium gems because of their exotic pedigree.

2

Pearls may have turned men's eyes toward the Upper Mississippi's mussel fishery in the early 1890s, but it was a German immigrant who really launched the shelling fleet.

John Boepple had been born near Hamburg in 1854. Like his father, he became a "turner"—a curious trade that involved cutting blanks from shells, animal horns, bones, and other materials. These would be refined into such gimcracks as stickpins, collar buttons, shoe buckles, and watch fobs. The work and the young craftsman suited each other; he was good at what he did and tried to get better, experimenting with various techniques and materials.

He had already worked with seashells of various kinds when someone in the United States happened to send his father a box of common river shells. He apparently gave them to his son, who found they could be readily worked, and he used the shells to make buttons and ornaments in his spare time. The young artisan was mastering his trade, soon opening his own shop and marrying, and the episode with the American river shells might have been forgotten if he hadn't been shadowed by tragedy. His young wife died suddenly, and at about the same time a stiff new tariff was imposed on all imported materials. Rising costs quickly closed Boepple's profit margin and it wasn't long until he was forced to close his shop.

Wife and business gone, Boepple took stock of his prospects and remembered his experiment with American river shells and the letter that had accompanied them explaining that such shells were plentiful and free to anyone wanting them. In 1887 he left Germany forever and headed for the American Midwest—land of milk and honey and free mussel shells.

A stranger in a strange land, speaking no English and knowing nothing about the new country, Boepple worked his way westward as a laborer, finally reaching Petersburg, Illinois, and the home of a sister who had immigrated earlier.

It was here that he got to know his first American river, going out to the nearby Sangamon to swim. And it was here one day that he cut his foot on what proved to be a sharp mussel shell; in fact, the bottom of that stretch of the Sangamon proved to be almost paved with shells. Boepple was enormously excited with the discovery, although his delight was tempered by certain realities. "At last I found what I had been looking for," he later wrote, "yet there still was a problem for me. I was without capital in a strange land among strange people and unfamiliar with the language."

As it turned out, most of the Sangamon River shell was too thin and fragile to be suitable material for machine-cutting. So were those in the Rock River farther north. But Boepple kept looking, and at Muscatine, Iowa, he found an abundance of thick-shelled mussels in shallow water where they could be easily gathered. He began carving novelty items from this shell, and although it sparked some interest and modest sales in Muscatine stores, Boepple wasn't exactly taking the business world by storm. He still made his living by farm labor, although he continued to prospect for good shell and ways of using it.

About then, fate and Boepple's fortunes began to reverse themselves. A high protective tariff had wrecked his business in Germany; now a similar tariff would put him into a new business. In 1890 a tariff had been levied on many products imported into this country, and those just happened to include the ocean shell mother-of-pearl from which the most attractive and durable pearl buttons were made. Expensive to begin with, buttons of this imported shell soon cost as much as a penny each and it was said that buttons had grown so costly that church goers had begun putting them into collection plates. Boepple was soon singing the praises of ordinary river shell as a source of fine pearl buttons, but no one paid much attention to the strange foreigner with the thick accent until Boepple actually made some buttons of river shell and began showing them around Muscatine. They were, as might be expected, excellent buttons—as good as any made from ocean shell—and it wasn't long until Boepple found financial backing. On January 26, 1891, he began making buttons full-time with his homemade turning equipment. A Muscatine basement became the first freshwater pearl button factory in the world.

Boepple's old-world work ethic began to pay off. Within a few years he had two hundred employees in his two-story brick factory, was earning a national reputation as the trailblazer in a colorful new industry—and was attracting the attention of the local business community.

As his production increased and his work force grew, it was impossible for Boepple to keep any secrets. Unless he wanted to return to a one-man operation, he had to share some of his cherished, hard-won technology with certain employees. It didn't take the local business hawks long to realize this. They began raiding Boepple's work force, hiring key people for better wages and more responsible jobs in new button factories—in return for the skill and information they brought with them. Some Boepple-trained button makers also went out on their own.

By 1897 there were three button factories in Muscatine alone, to say nothing of ones springing up in river towns as far away as Hannibal, Missouri. The entrepreneurs were in the game now, the sharp money men with ideas for automation and mass production that John Boepple had never dreamed of. It was a rich new market that had struck down any possibility of foreign competition; ocean mother-of-pearl now cost as much as sixty dollars per hundred pounds, and the costliest of the freshwater mother-of-pearl was about one hundred times less—at about sixty cents per hundred pounds. The domestic pearl-button market was booming, as were international sales, and the freshwater pearl button actually became a significant item in our balance of trade. But as the strange new industry prospered, John Boepple began to fade. Clinging to his ideals of craftsmanship and hand-wrought quality, he was lost in this new world of mass production, automation, and high-powered marketing techniques. He remained a rather quaint anachronism with a thick accent and inflexible standards, still a stranger in a strange land and a stranger even in the burgeoning

industry he had founded. His backers soon shunted him off to a new branch plant in nearby Davenport—and he had no sooner left Muscatine than his home plant was converted to automatic production.

From then on it was mostly downhill for Boepple. He became a buyer for other shell companies, working for a time in Wisconsin and Indiana, and was finally hired by the Fairport Biological Station in Iowa to help study life cycles of the freshwater mussels that had been so greatly depleted by the industry he had sired. His career ended with one of those ironic flukes that had marked so much of his life. In 1911, while wading in an Indiana river sampling mussels, he again cut his foot on a shell. It became infected, and blood poisoning developed. Boepple returned to Muscatine where his condition grew worse. He went into a coma and died in a Muscatine hospital in January 1912.

While it lasted, the button industry of the Upper Mississippi was a wild and colorful free-for-all that made some men rich, others poor, and some dead, and made everyone take a new look at the River. For fifty years Muscatine was the center of it, the Pearl City at whose front door lay a great bed of mussels being worked by a fleet of three hundred fishing boats. Whole families took to the River; every summer large clamming camps sprang up in northeastern Iowa and Wisconsin and Minnesota, and packet boats plied the River selling groceries and other supplies at swollen prices. Farmers complained as sons and hired men deserted farm drudgery for the color and noise and quick money of the clam camps. Riverfront saloons boomed and riverbanks were piled high with ripening clam meats. The wild islands became even wilder and some camps were virtually beyond the law. Bearded, sun-blackened rivermen came upstream from Cape Girardeau and downstream from Lake Pepin and the Wisconsin and Minnesota timberlands. They worked for a time, blew their shell money on roaring sprees, and disappeared. Often someone else's boat or pearls vanished with them.

I can think of no one who has a better handle on this wild and little-known side channel of American history than Dr. Mike O'Hara, chairman of the Department of Composition and Literature at Muscatine Community College. In a paper given at a 1980 symposium on Upper Mississippi River mollusks, Mike told of actual traffic jams over the big clam beds with boats stretched from shore to shore—and in such cases there might be minor naval clashes. A pearler from Alabama went so far as to mount a small cannon on his houseboat, which was anchored on a mussel bed to which the sheller claimed sole rights. One day a Chinese sheller began a drag over that particular mussel bed (probably not understanding English) and was promptly blown out of the water. From then on, no one disputed the man's claim. He had no legal basis, of course, since one doesn't file mussel-bed claims in the same sense as gold claims. But even a little cannon can spout a lot of instant statute.

The river towns have never been as tame as those even a little farther inland, and as the shelling industry heated up so did its attendant culture. Fights and street crimes were rampant, and Mike O'Hara notes that if a sheller found a good pearl he was wise to put it into his mouth under his tongue—which not only hid the pearl but kept the finder from bragging about it until a buyer came along. There was the predictable amount of flimflam, with sellers approaching pearl buyers at night and trying to palm off marbles or bits of polished anthracite coal as pearls. "In spots where pearls had been found, townspeople might line the banks to watch the clammers," O'Hara reported. "Some smart clammers would bring their catch ashore and sell it unopened to those spectators."

By 1898 there were forty-nine button factories in thirteen cities on the Upper Mississippi and a dozen others along other rivers. And the pristine mussel beds were ravished apace. In 1896—several years before the apogee of the button boom—five hundred tons of shell had been taken from a single mussel bed two miles long and a quarter-mile wide. Two years later a thousand full-time fishermen were working 167 miles of river between Fort Madison and Sabula, halfway up Iowa's eastern border. A bed one and one-half miles long near New Boston, Illinois, yielded ten thousand tons of shell in three years—a probable total of a hundred million mussels. As the great mussel beds of Muscatine and southeastern Iowa petered out, the shelling extended southward into Missouri and upriver into Wisconsin and Minnesota. In 1899 it was the most valuable fishery in Wisconsin and yielded over sixteen million pounds of shell.

The first of the clamming was done in late summer until about freeze-up—say, from August into December—probably because river levels were more moderate then. Later on, the clamming was done year-round, with actual ice fishing beginning in the winter of 1896. Some unsung genius introduced crowfoot bars in the spring of 1897—a significant technological breakthrough, since a crowfoot bar probably cost no more than two dollars at the time and wasn't difficult to use. With full winter, clamming was done through holes in the ice with several types of rakes. One man could rake almost a half-ton of shells in a day, hauling them over the ice in boxes on sleds. Tough work, but so were the men doing it. Such winter fishing grew in favor for it was said that the quality of winter shell was better, being less brittle.

The shelling proceeded, unregulated and with growing intensity and efficiency. It did not take long to skim off the cream; at the beginning, a man could earn thirty dollars a week or even more working on a productive bed of mussels. During 1899 the average was much less than that, and was probably below ten dollars a week. But that was at a time when working men were lucky to make a wage of a dollar a day. Besides, a day laborer or farmhand ashore didn't stand much chance of finding a fine pearl.

By 1912 there were two hundred button-making plants in the United States, with total sales of more than six million dollars a year. The sale of freshwater pearls and baroques alone amounted to about three hundred thousand dollars annually. The industry continued to grow through World War I and into the early 1920s, with as much as sixty thousand tons of raw shell being taken yearly and selling for at least a million dollars.

No shellfish can withstand such immense fishing pressure, yet it continued even when the markets were glutted with raw shell and only one-fourth of a man's catch might be sold. The great Muscatine beds began hurting within ten years after the first button factory appeared, and the decline spread upstream and down. Even Minnesota's mighty Lake Pepin—that inexhaustible fountainhead of mussels—was being stripped. In 1914 it had produced eight million pounds of marketable shell, but fifteen years later the total catch was only 4 percent of that. By 1946 there was no shelling in the Mississippi below Muscatine at all, and even during ideal weather and river stages there were no more than a dozen clamming boats on the Muscatine waterfront. In the late 1950s I knew of no more than four part-time shellers at Harper's Ferry.

Mussels grow slowly and conditions must be right for their increase. Females discharge huge numbers of larvae, their *glochidia*, into the water and these tiny organisms must be adopted by another animal or die. They pass through a parasitic phase in which they may encyst themselves in the gill filaments of specific fish. There they grow for a short time before leaving their hosts and falling to the river bottom. Any surviving this chancey part of their life cycle must still survive for at least ten or fifteen years before reaching commercial size. A shell four-and-one-half inches in diameter may be eighteen years old. The most valuable of the early commercial mussels, the "niggerhead," might require a dozen years to qualify as a three-inch marketable shell. And even as the first signs of overfishing began to appear, scientists of the day thought it unlikely that any mussel bed could replenish itself while being commercially productive.

Early attempts were made to save the mussels and the valuable industry they fed. New laws were enacted to greatly curtail shelling, although enforcement was spotty and weak at best. The United States Bureau of Fisheries set up a biological station at Fairport, Iowa, in 1908 and began artificial propagation of mussels. Young clams were produced in vast numbers in darkened troughs and host fish such as sheepshead and buffalo were infected with the mussel larvae and released. During only two months in 1920, rescue crews infected six million fish with nearly half a billion *glochidia*. It may have helped a little, but not enough to revive a dying industry. The primeval mussel beds had been ravished far more rapidly than they could ever replenish themselves. And even as the heavily farmed watersheds poured eroding topsoil into the Upper Mississippi and many of the famous old clam beds were being buried by silt that gathered behind the newly created channel dams of the 1930s, and pollution from growing cities added its venoms, the fishing pressure continued.

The *coup de grace* to the dying industry was dealt shortly after World War II with the development of whole new families of plastics. Buttons of mother-of-pearl could not compete with new plastic buttons that were more easily and cheaply made and far tougher than pearl buttons. I can't remember breaking a shirt button in the past thirty years, but it was a common nuisance before then. The new plastic buttons were said to endure detergents and changing laundering methods better than mother-of-pearl, even though some of the old pearl button display cards proudly announced that "They Wash."

Politics and technology had provided a commercial use for the mussels of the Upper Mississippi. Politics and technology—in the guise of riparian factories, agribusiness, channel dams, plastic buttons, and rampant exploitation—had decimated a resource base and expunged an industry. The System giveth and the System taketh away. And although it never occurred to us that mist-shrouded morning on Harper's Slough in 1953, Joe Martelle and I were dragging crowfoot bars through the end of an era.

3

My first nonacademic instruction in the great class *Pelecypoda* occurred in the sandy shallows of the Des Moines River in central Iowa, just downstream from the low-head dam where Dad was fishing for crappies and growling about nine-year-old boys who lacked the proper stick-to-itiveness.

But the fishing was slow, and the broad sandbars beckoned. I still recall it, over fifty years later, as one of those golden days fabricated with kids in mind. The sand was gold, with a golden shade to the perfectly clear water that ran across the head of the main sandbar and just behind it, breaking up into a maze of shallow chutes and pools, with the golden afternoon in full flood and anointing me with a full crop of golden freckles. There was a little backchannel, maybe ten yards across and no more than a foot deep, with a perfectly smooth sand floor—except where it bore odd furrows that looped and bent, often crossing themselves: curious marks that started and ended suddenly, sometimes grooving the sand bottom for ten or twelve feet. They were like nothing I had ever really noticed before, and I didn't pay them any close interest until I found an almost-buried mussel at the end of one. The big clam was hinge-edge up and in the act of burying itself after traveling from one featureless patch of sand to another. As I pounced on it, I knew I'd made one of those great breakthroughs that have illuminated the dark journey of humankind. Clams I knew something about; at least, I had seen their empty shells. But I'd never caught one alive—and in the act of moving. Those mysterious furrows were *trails,* with an actual critter buried at one end or the other! The already fine afternoon took on a rich new dimension, infused with purpose and thereby achieving perfection, and I ranged through those shallows digging like a freckled raccoon. There was all manner of mussels buried at those trail-ends, and even today, allowing for the magnification of memory, I think those were unusually large ones. They were big, heavy, and thick—probably muckets or washboards—and very probably representing an old population, since I have never heard of any commercial shelling on that stretch of river.

Anyway, I took some of the largest ones and we hauled them home in Dad's bait bucket and the old zinc-lined cold chest, and I kept them in the backyard in a half-filled washtub. They drew kids from all corners of the neighborhood and beyond, but after a few days in the sun the mussels "join'd the choir invisible" and became much less attractive. In fact, they became singularly unattractive. At about this point some grown-up informed me that such large clams often held large pearls—any one of which would provide a lifetime supply of cherry phosphates and cowboy guns. Those mussels didn't even require opening; they had already fallen open by themselves and were gaping in a well-advanced state of putrefaction when I began treasure-hunting in their remains. There were no pearls. There was, however, a distinctive air about them that has outdone all other airs before or since and is the first thing that comes to mind when someone asks me if I have ever eaten freshwater mussels.

A long time after that I learned that those mussels were probably males sowing sperm along the furrows being plowed in the fine sand of the river shallows. Or perhaps they were just blindly seeking some outlet into deeper water away from the light and heat of the full sun, to where stronger current would sweep more food and oxygen to them. Strange and varied animals, these freshwater mussels, with a wide variety of common names conferred by a wide variety of river folk. Consider such *nomina vulgaria* as three-ridge, sandshell, pocketbook, heelsplitter, wartyback, pigtoe, pimple back, washboard, elephant's ear, buckhorn, spectacle case, sheepnose, mucket, along with such flowery offerings as the orange-footed pearly mussel. The diversity of species in the Upper Mississippi (there are thirty-two species in Lake Pepin alone, where as many as fifty mussels per square yard have been found) has led some scientists to believe it was there that freshwater mussels first developed in the New World.

All these are "naiad mussels" that discharge clouds of the parasitic *glochidia* that encyst themselves in some part of a fish, with hookless species usually ending up in the gill filaments and the hooked types of *glochidia* attaching themselves to fins or other parts of the body surface. In contrast, sea-living bivalves produce free-swimming larvae.

Since *glochidia* usually die in only a few days if contact with a host is not made, there are ways and means of assuring such contact. The tiny *glochidia* may be discharged by the millions from larger species of mussels, drifting for a time in temporary suspension before sinking to the bottom. They may be inhaled by a fish and passed into its gills or stirred up from the bottom by fin movements or as a fish feeds. Or, consider the pocketbook mussel, which develops a fleshy papilla at spawning time—a minnow-shaped structure that even has a slowly moving "tail fin." Any fish attracted by this decoy will be bombarded with a cloud of *glochidia*. Infections by these tiny parasites rarely injure the host even though up to three thousand *glochidia* have been found encysted in a single fish. This parasitic stage may last as long as six months, but it usually extends from ten to thirty days. Then the young mussel breaks out of its encystment and falls to the stream bottom as a juvenile, not reaching sexual maturity for at least a year, and as much as eight years in some species.

Most types of naiad mussels are not particularly fussy about their host fish, and some are capable of infecting many species. Other mussels are intensely host-specific, and the deadly danger of such specialization is exemplified by the ebony shell, whose host is the skipjack herring. Seventy years ago the ebony was a common species in the Upper Mississippi, and the skipjack was allowed to pursue its annual migrations unimpeded by any major obstacles in the River. Then came the great channel dams of the 1930s, and the skipjack's wandering ways came to an end. So, for all practical purposes, did the ebony shell. A few ebonies are still found—especially in the River just above Muscatine—but only as very old individuals. As long ago as 1931, biologists were unable to find an ebony shell in the Upper Mississippi that was less than nine years old. Since that was over fifty years ago, it is possible that the youngest ebony shells in that part of the River today may be sixty years old.

About the same thing is true of the yellow sandshell, even though that mussel was a generalist whose *glochidia* were at home in at least nine host species of fish. There were broader forces at work—changes in bottom composition, deteriorating water quality, and a whole host of nasty little factors that are scarcely understood.

The most direct assault on the River's mussel populations is the actual shellfishing, of course, but a close second is dredging to maintain the navigation channel or as commercial sand-and-gravel operations. Channel maintenance is not that bad if it is the one-stage type in which the spoil is discharged up onto dry land where it cannot find its way back into the River. But two-stage dredging is something else. This is done when the spoil banks are some distance from the channel and the sand is discharged into inshore shallows and then re-dredged to be dumped farther inland. There probably wasn't much good mussel habitat in the shifting sands of the main channel anyway, for the factors that made dredging necessary were also inimical to mussel concentrations. But those inshore shallows where the second stage of dredging occurs are another matter; these may be prime mussel habitats that are buried by the first stage of dredging and torn away by the second. Dredging for construction materials is likely to occur also in the fairly stable shallows outside the thread of the main channel, destroying mussel beds that may have existed there for centuries. No one knows how old some of those beds are, but there are places where the layers of ancient shell are several feet deep.

As some beds are destroyed by removal of bottom materials, others have been buried forever by sand and silt washing down from eroding uplands where stabilizing plant cover has been stripped away by modern agribusiness. There was a time when those materials might have been generally flushed downriver by the same freshets that carried them into the Mississippi, but this changed with modern channel engineering—and the advent of wing dams and the big channel dams. These are effective silt traps, burying certain mussel beds even as shifting bottom patterns have scoured away others.

Some of the same works that wreck mussel-bedding environments are also degrading essential water quality. The upland erosion that pours sand and silt into the River brings biocides of many kinds, heavy loads of fertilizer and organic wastes, and the particular poisons that emanate from cities as sewage and industrial effluents. As the River's load of organic pollutants has risen, so have certain microorganisms that can infect mussels and cause sterility. The mussel population from the Twin Cities to Lake Pepin is only a wan shadow of what it was, and from Saint Louis southward, native naiad mussels in the Mississippi River have been so greatly reduced that for years it was thought there were none at all.

The scanty mussel populations below the Twin Cities and the mouth of the badly polluted Minnesota River persist to a point well below the entry of the Chippewa River. The Chippewa isn't a poisoner of the Minnesota's caliber, but it carries an immense burden of sand that maintains the natural blockage at the lower end of Lake Pepin and creates a shifting streambed that just isn't good mussel country. Not far below the Chippewa, though, the River enters a "recovery zone" in which the mussel fauna begins a revitalization that continues on downstream to about the mouth of the Iowa River not far below the old shelling center of Muscatine. Something happens below the mouths of the Iowa and Des Moines rivers—possibly the same thing that happens just below the mouth of the Minnesota. Whatever it is, the mussel fauna declines steadily from southeastern Iowa downstream.

There is still some commercial shelling as far south as Pool 26 at Alton, and in the lower reaches of the Illinois River—which is even more badly abused than the Mississippi, but recent surveys indicate that mussel populations in the last six pools of the Upper

Mississippi and in the free-flowing Middle Mississippi from Saint Louis to the mouth of the Ohio are in a bad way.

Although today's mussel beds in the Upper River are only dwindling relics of what they were, and in spite of the plastic buttons, pollution, habitat destruction, and almost ruinous exploitation, commercial shelling never really stopped. There was a market of sorts even after glass and plastic buttons came into almost universal use during the 1950s, and a handful of die-hard rivermen like Joe Martelle bridged the gap between the pearl button of yesterday and the cultured gem pearl of today.

In 1896—the year before the crowfoot bar was first developed and the Upper Mississippi's pearl-button industry began to really take off—a Japanese named Kokichi Mikimoto succeeded in culturing small blister pearls in pearl oysters. Pearl culturing was not new; the Chinese had experimented with it a thousand years earlier, and the Swedish botanist Carolus Linnaeus had produced cultured pearls in freshwater mussels in the eighteenth century. But it would be the Japanese who persisted most doggedly, finally developing ways to create cultured pearls under carefully controlled conditions with the use of mother-of-pearl pellets cut from freshwater mussels.

As foreign agents that stimulate the secretion of protective nacre when inserted between the shell and fleshy mantle of a pearl oyster, such pellets have no equal. For one thing, they provide a pure mother-of-pearl nucleus for the thin layer of pearl oyster nacre, and even more important, they generally lack the mysterious "tolberg layer"—an organic wax to which pearl nacre will not adhere. If a pellet having a speck of this tolberg layer is introduced into an oyster, the pearl nacre will not be uniformly deposited and the result is likely to be an irregularly shaped baroque rather than the perfectly spherical pearl that is the darling of the gem trade. The ideal nucleus for a cultured pearl, then, is a pellet cut from freshwater mussels that are hard, thick, available in suitable quantity, and generally lacking in the tolberg layer. Those criteria are better met by musselshell found in such rivers as the Upper Mississippi, the Tennessee, and the Wabash than in any others in the world. For years the Chinese have tried to sell the Japanese pearl culturists their musselshell at a fraction of the going American price, but with no success, because about 90 percent of the shell from the Yangtze and other Chinese rivers contains tolberg layers.

So, beginning in about 1960, a fresh new market developed for the mussels of the Upper Mississippi. Prices for raw shell were higher than ever before, and the demand was for such large, thick shells as those of the three-ridges and washboards—species that were not only found in all the Upper River impoundments but were also in relatively good supply.

In Japan, the selected American shell is cut into strips that are then sectioned into small cubes. These are rounded off to form small pellets of mother-of-pearl several millimeters in diameter. The size of the pellet is contingent on the size of pearl it is expected to produce, for only about one millimeter of actual pearl nacre will be deposited. If an eight-millimeter pearl is desired, a pellet of about seven millimeters will be wrapped in a bit of mantle tissue and carefully inserted in the body of the oyster. Some novel specialties have developed; for instance, a tiny tin image of Buddha may be inserted in the pearl oyster—and a year or two later the oyster will be opened to reveal a little Buddha of pure pearl adhering to the shell.

By 1966 the export value of American mussel shells was nearly nine million dollars, which one exporter estimated to be the equivalent of thirty million dollars at today's price levels. Twelve years later, about seven million pounds of shell were exported from the United States; almost all of it went to Japan, where 90 percent was used as cultured-pearl nuclei at about three hundred "sea farms." Bankside prices for raw shell in 1966 rose to five hundred dollars per ton—and a new generation of shellers took to the rivers.

Some of them went the old way, with a pair of light crowfoot bars, a medium-sized boat, and lots of advice from granddad. A few of the *avant garde* literally took a technological plunge, diving for shell with the use of scuba or boat-based air pumps—an efficient, highly selective way of gathering commercial shell if you don't mind some of the attendant risks. I'll say this about it—the diver isn't in the hot sun. There isn't even much light in his dim world of brown murk, and he is seldom able to see his hand when it is held against his face mask. He works by touch, groping along the bottom for big shell, often working near the sides of the main channel within a stone's throw of passing towboats whose great "wheels" could convert him into an instant sackful of hamburger and torn neoprene. He is probably discomfited by the ear infections that plague such river divers—the least of the occupational hazards he must face in that dark-brown world where he can never see the dangers that beset him.

Not long ago I met a young man—hardly more than a boy by my measure—who is already a veteran shell diver in the Upper Mississippi north of Dubuque. One day, diving free with scuba, he went down near the bank on the channel side where the river was deepest and found himself in a situation he could not understand. He seemed to be trapped in a large compartment of some kind, and in the darkness he had trouble finding his way out. When he finally did, he realized he had dived down through the open door of a sunken railroad boxcar that no one had known was there. It was most emphatically not a good place to die; since then he and his partners have always worked with lifelines.

But for all its built-in hazards, diving for shell gives the fisherman an edge that the topwater worker doesn't have. The father of one of the divers engaged in that sunken boxcar episode told me that his son and two other boys, working the seams of mussels south of Prairie du Chien, sometimes filled their boat with shell in a little more than four hours, making as much as fifteen hundred dollars a week. Those mussels were shipped alive, and the boys were spared the onerous chore of "cooking out." The Japanese, he told me, "wanted the meats for medical extracts" of some kind.

High shell prices, with a whiff of adventure and the chance to turn a dollar working under water, have lured divers to the Upper Mississippi from as far away as the east coast and deep South. The process is simple enough. A diver works from a boat with scuba or an air hose connected to a compressor—a "hookah," in diving parlance. If working with scuba the diver must have a workday's supply of air—enough tanks for six to ten hours of diving. The boat is solidly anchored while the diver works the mussel bed in the immediate area, crawling over the bottom on hands and knees, dragging the wire fish basket and groping for shell in the near-darkness.

There are some bitter feelings about this sort of thing. I've spoken with old-time shellers who waxed downright incandescent about the destruction wrought by divers, and the feeling is shared by some fisheries biologists. In a noteworthy 1980 symposium on Upper Mississippi mussels, malacologist Samuel L. H. Fuller said that "hundreds of tons (at a minimum) have been wiped out by scuba divers, which are widely used in the modern fishery." Another side is voiced by certain commercial shellers whose experience includes scuba as well as the more traditional gear. Worth Emmanuel of Maiden Rock, Wisconsin, is one: "I am told that divers totally destroyed a bed at one time. Depending on the interpretation, I suppose this is possible. For instance, when divers work a bed the percentage of large shell left behind is so small that it is not practical to again work it commercially for two to five years or until regrowth brings the shell up to size again . . . but it is impossible for the diver to get all the large shells, let alone all the shells. I feel it is impossible to wipe out a bed in the literal sense." Agreeing with him is John Latendresse of Camden, Tennessee, who probably knows as much about the shelling industry as any man alive. He feels that divers can "completely deplete" populations of mussels in clear water of high visibility but does not believe it is possible to take more than half a mussel population by diving in waters of such limited visibility as the Illinois and Mississippi. I wouldn't know about that—but I have done my share of free diving in murky waters and have had trouble finding tackle boxes and shotguns even in places where we had exact fixes on the lost objects. A few weeks before I wrote this, experienced scuba divers were unable to find a helicopter that had crashed just off the Saint Louis waterfront, and it's a bit hard to believe that they could have wiped out a bed of river mussels under the same conditions. Still, there is an official ban on this kind of clamming in some areas; it is now unlawful to dive commercially for mussels in waters of Ohio and Kentucky. It is hard to say whether such regulations are based largely on biological evidence of mussel depletion caused by divers, or as a result of political pressure exerted by traditional shellers who buy far more commercial fishing licenses than do divers. On the other hand, some divers strike back with the argument that their method of highly selective mussel fishing does not disrupt the beds or injure and displace undersized mussels as do crowfoot bars, rakes, and shell dredges.

The old traditional ways of shelling have also been brought up to date, with store-bought crows' feet, plastic ropes, and—glory be! gasoline-powered winches for raising crowfoot bars. Even so, it doesn't take an awesome investment to get into the commercial shell game, and during the recent recession—or whatever it was, depending on your political bent—some unemployed citizens chose to go on the River instead of the dole.

I drove past the Harper's Ferry riverfront early one summer afternoon, the first time in years I had been there, expecting to be disappointed. And was, of course. I can usually count on being disappointed if I expect to be. The place had been thoroughly cleaned up, civilized, and put in order. The little old commercial fishermen's shacks were gone, with their net racks and cook-out tanks, rickety docks and the inevitable wooden boat peacefully rotting in retirement. The place didn't even smell the same. It smelled of charcoal grills and barbeque, a contribution of the new mobile homes and neat weekend cottages that now fronted Harper's Slough. The scruffy, cruddy old riverfront had been sanitized; all it needed was a paper strip across it, like the ones across toilets in motels, certifying sterility of place and function. Everything was neat, clean, and in order. No smells of dead fish, hot tar, sweat, woodsmoke, or wet dogs. I sat on a patch of neatly mowed grass and felt sorry for myself. Joe Martelle was gone; so were Jim Williams and Three-Finger George Kaufman. I brooded. From somewhere nearby came the buzzing rattle of a window air conditioner. And why not? After all, hadn't the temperature soared to a blistering eighty degrees? It all fitted in with the water skis and the candy-assed twelve-thousand-dollar bass boat over there in the driveway.

But as I stood to leave, an apparition came up Harper's Slough— a beat-up wooden workboat, broad of beam and blunt of bow, with a broad, blunt man at the helm. Crowfoot bars racked on each gun'l, with a cargo of at least fifteen buckets of shell amidships. He docked at a little landing near a battered pickup truck and began putting things in order, and my day brightened forthwith. He was wearing disreputable pants stained with oil and river mud, a filthy shirt, and an old red hunting cap with holes cut for ventilation. This was Carl Rider of Prairie du Chien, and he'd been having some pretty good hauls of three-ridges and washboards on Harper's Slough.

"You been at this for quite a while?" I asked him after we'd howdied and observed the necessary courtesies that are always prefatory for opening a casual conversation on the River.

"Not for long," he replied. "Not full-time, anyway. My real game is the construction business. Been in construction for years, but it fell on its ass and a man's gotta do something. Why hang around town lookin' for work that ain't there? Instead of doin' that I took to the river and went to clamming. It's about my fourth start in life, but a man oughta start fresh once in a while. My kids are all raised and educated, and a while back my wife got her degree in nursing, so she's got plenty to keep her busy. So here I am on the river."

"You been seeing any divers working out there?"

"Not for the past week or so," Rider answered, never stopping work. "There's been some working up above Lansing, I hear. Been some around here and down south of Prairie, too. You know, two of them young divers come all the way out here from Maryland? Yeah! Just to work these beds. I'll bet they know what they're doin', too. Some of these divers ain't qualified at all, you know. Ain't that a helluva note?" He shook his head despairingly, still working steadily. "Diving right out there by the main channel, and not even qualified!"

"How are the beds around here holding up? Still quite a bit of shell?"

"Plenty as far as I can tell. I've heard that some core samples were taken out there—by the Engineers or somebody—and they found old shell down to thirty or forty feet. There's live shell on top of all that, too."

"Always more where that came from, huh? What do you suppose the scuba divers will do to it?"

"Won't do it any good," Carl said without rancor. "Divers can hit a bed real heavy, and they never stir it or rake it like a bar does. I think you could take shell almost indefinitely with crowfoot bars and not wipe out a bed. But scuba—I don't know. Say, next time you're coming up on this stretch of the river let me know, and go out with me. It's a helluva job for one man alone, even with this gas winch. I might be working farther downriver than this. They'll know at the Tennessee Shell Company's buying station at Prairie, just south of the foot of the bridge. I mean it. I'll welcome the company. You take it easy."

I left town feeling better than I had all day. Part of the Harper's Ferry I once knew was still hanging on, still alive and well down there by the slough, even though things would never quite be the same again without Katie Quillan's apple pie.

But I missed the pearl talk, up and down the River. It had been a long time since I'd seen anyone light up with stories about big pearls and their finding, with mysterious hints about an old lost seam of shell that granddad had hit one day when he was drinking and could never find again, with its clams like pie plates and pearls like banty eggs.

Then came the lowering day when I put in at Noble's Landing south of Harper's, my mood as overcast as the sky. I had camped on a flat beside Paint Creek the night before and had sweltered in breathless humidity that had drenched my tent and dampened the sleeping bag, and now the new day promised rain and no chance to dry my outfit. When I launched the boat the motor was running rough and I spent an hour cleaning spark plugs and messing with a stubborn needle valve, remembering words I hadn't used since high school. But things picked up as they always do on the River if you give them a chance, and my motor started to smooth out just about the time a shelling boat came by, headed downstream.

It was a capable looking outfit, a big wooden johnboat designed for commercial work with the gas-powered winch you'll see on today's serious shelling boats. It was crewed by two young men, scarcely more than boys but already with that indelible river look to them, that indegoddampendent Huck-Finnish aspect that validates the pedigree of a born-and-bred river rat. I waved a pair of pliers at them and as they waved back the fitness of the scene struck me. They were passing the southern end of Jackson Island. Now, there is no Jackson Island across the river from Hannibal, Missouri. Mark Twain made that up, and had Tom Sawyer and his buddies playing pirate on an island that never existed. There are

Shuck's Island and Glasscox Island and King's Island there, but no Jackson Island, no matter how hard the tourists look for it. The real Jackson Island was here, just above the mouth of Paint Creek, and so were a latter-day Huck and Tom—a bit older than when Mark Twain left them—out on the river to "bar a bed" in the gray light of a cloudy morning.

They were out of sight by the time my outboard motor was properly adjusted, but it didn't take me long to catch up with them. They were anchored near the edge of the channel working on something in the boat, and hailed me as I passed. I turned back and drifted down on them."

"Got problems?"

"Sure have," said Raymond Boland. "You got a screwdriver we could use?"

I produced a screwdriver. "Anything I can do?"

"Naw. Busted recoil spring on this old Briggs and Stratton engine that works the winch. Another guy used this boat a couple days ago and busted this starter and didn't even tell me. On top of that, he took the screwdriver and knife that I always keep here in the boat. You wouldn't have a knife on you, would you?" I produced a knife. The problem was analyzed from several angles and it was opined that the only solution at hand was to mount the starter-cord pulley directly on the starter shaft, dispensing with the recoil mechanism altogether. It was further opined that whoever yanked that starter rope had better be ready to let go of it in one helluva hurry. This plan was executed flawlessly; the engine started on the second pull, Raymond's arm wasn't broken in the process, and I took my leave and headed downriver.

I had important business in a marshy backwater south of Prairie du Chien, where a very big black bass frequents a certain bed of lily pads. Either the bass wasn't home or had already breakfasted, for my most earnest blandishments elicited no action. After a couple of hours I had finished my thermos of coffee and seen as many cow lilies as I cared to see, so I poled back out into open water and cranked up the outfit. It was beginning to rain steadily now; I donned oilskins and returned to the main channel, driving through shifting curtains of a surprisingly cool summer rain. Twenty minutes later the big wooden johnboat appeared ahead, drifting toward

me with one crowfoot bar working and the other being picked. Ray
and his partner worked oblivious to the rain, thin clothing plastered
to their bodies. No raingear for the spiritual heirs of Tom and Huck.
Not in high summer, even if there was a slight chill in the wet
wind. I cut power and took up my oars, keeping pace with the
drifting sheller.

"Doing any good?" The age-old question.

"Naw. It ain't much good today for some reason. We'll prob'ly
only make a few hundred pounds."

It was time to lift a bar again; the little engine was cranked and
the winch engaged and the heavy crowfoot bar emerged dripping
from the river, clusters of clams hanging from its chains. With the
other bar dragging, they began picking shell. Many undersized
mussels were thrown back when they were found to pass through
two holes in a wooden template that sized the legal shell. One hole
measured the minimum legal size for washboards and the smaller
hole determined legal three-ridges. "They're the only kinds we're
taking," Raymond explained. "Three-ridges and washboards and
once in a while a big wartyback."

"Slow day, huh?"

"Oh, hell, yes! One day earlier this year we took two thousand
pounds. One day last year we took three thousand. Now, there's
a pretty good day."

"Today wasn't hardly worth coming out on the river, then."

"Well, it's *something*," said Raymond. "Beats setting around all day in some tavern, and I can't afford that, anyway. If I ain't doing this I'll prob'ly be commercial fishing. I was doing some earlier, but hung it up when the market started to drop. It's come up again, though. Buffalo are back up to about twenty-three cents so I may go back to fishing if I don't do no better than I am clamming."

A big tow was bearing down on us so I shipped oars and started the motor and swung over toward the Wisconsin shore, wondering just how many times in the course of a year commercial fishermen on the Mississippi have to pull their gear and yield to the big towboats. In theory, they have just as much right to be out there on the River earning a living as the big barge lines. I also wondered if anyone ever really believes things like that. If anyone does, I have a couple of big bridges I'd like to sell them.

Running toward the Wisconsin shore I noticed an odd white patch just above the water's edge. It proved to be a bank of very old shell, obviously left there by some sheller long ago. I couldn't explain this, and when I rejoined Raymond neither could he.

"It's pretty good-sized shell," he said. "Don't see why anybody would go to the trouble of catching it and just throw it away. Maybe dad will know."

"Your dad a commercial fisherman?"

"Naw. Logger. But things are pretty slow at the mill and so he comes out and helps me when he can. I think he'd rather fish than clam, though. He sure dotes on catfish livers. Really gluts himself early in the season."

For some reason the talk began to turn to fall and squirrel-hunting, followed a natural progression through duck-hunting and deer-hunting, and then led farther afield than I'd have ever expected. It developed that Raymond had actually hunted moose up in Ontario, having gone with some local men a couple of years before. Moose-hunting is a disease, and a very expensive one. I told Raymond so, as if he didn't know it.

"Oh, I know it all right," he replied. "But damned if I ain't going this fall if I sell my pearl."

"If you *what*?"

"I found me a real good pearl last year," he went on. "A guy over at Cedar Rapids wants to buy it, and we're dickering. If I get my price, and I think I will, it ought to get me up there and buy my license and pay for a pretty good moose hunt."

He had my undivided attention. Your basic Ontario moose hunt today is bound to cost at least two thousand dollars.

"You must have yourself a pretty good pearl."

"I have," said young Raymond Boland.

I haven't been back to Waukon Junction and seen him since then, and I don't know if he ever parlayed his sweetwater pearl into a Canadian moose hunt. I hope he did. The few Tom Sawyers left in this world deserve a treasure now and then, and any adventures that go with it.

There is a bivalve found in the Upper Mississippi called the Higgin's eye pearly mussel. Its Sunday name is *Lampsilis higginsi* (Lea), and the primary host for its *glochidia* is the sauger pike. The Higgin's eye seems to prefer the somewhat deeper parts of big rivers in places of high turbulence and oxygen content, and it is apparently found nowhere else in the world but in the drainage of the Upper Mississippi north of the mouth of the Missouri. Any species of naiad mussel that exists in only a single river system in this day and age is walking a biological tightrope and the Higgin's eye is no exception. It is a recent inductee into the sad society of endangered species.

This was noted in our local newspaper not long ago, and a reader wrote to the editor: "So what if things like the snail darter, furbish lousewort, and Higgin's eye mussel vanish forever? They didn't do us any good when they were here; what possible difference can it make if they're gone?"

I've been wondering that, myself. And if we can't figure out what something is good for, why not let it slip away into oblivion? Who cares, indeed?

But then, what is modern man good for? Of all earth's creatures he is unique in not being good for anything. He is equally unique in being bad for almost everything. No other critter can make that claim. The sad truth is, our lovely little planet can no longer afford us. We are Earth's only bad habit, one that started out harmless enough a few million years back, but which has become a major vice.

Thinking of myself as a bad habit that's gotten out of hand isn't the sunniest of outlooks, nor is the realization that the Higgin's eye mussel contributes more to Earth than I do. The mussel helps stabilize the streambed in which it lives; it filters and clarifies water that passes through it, straining out suspended materials and converting tiny organisms to tissue that can be used, in turn, by such higher forms as fish, otters, muskrats, waterfowl, and crawdads. I can't do those things; I tear at the riverbeds, poison the food chain, and corrupt the waters that sustain me. All that is bad enough. But the real blow to my lordly pride is the knowledge that while the mussel can make a pearl, the best I can do is gallstones.

3

JOHN A. GRINDLE AND ASSOCIATES

False dawn, the gray quarter-hour that comes just before the crow and after the owl, was as cool as the early September day was going to get. I was trying to get deeper into the light sleeping bag but some sort of obstruction was giving me trouble. I raised my head to see a black Labrador retriever lying on the foot of the bag.

"Hey! Get off the bedding!"

The big dog lazily opened his larboard eye and thumped the bag with his tail. I cocked one leg inside the sleeping bag and pushed hard, rolling him off onto the sand where he lay on his back with all four feet in the air, looking at me upside down with his ears spread out and a silly grin on his face. He rolled over and stood and shook himself, spraying sand widely, then walked around and sat near my head, still grinning.

"You know what I do with worthless black dogs like you? I *grab* 'em, and set on 'em, and then cut 'em up into little pieces for catfish bait."

Thumpthumpthump said the tail.

"But first I'm gonna cut off that black otter tail of yours and beat on you with it! Ever since you were in that hunting movie, you've thought you're Rin-Tin-Tin!"

Thumpthump, which translated roughly as well-now-that-you're-awake-let's-do-something. Even under the bludgeon of my wrath, the silly grin never faded. No respect at all. That's what you get when you cross a perfectly good American swamp dog with a blonde English bitch of noble line. No respect for us mongrels.

The island camp was a good one, with huge silver maples bending out over bare sand and all the poison ivy growing farther back on higher ground. Tar and I had stopped there in late afternoon with more than enough time to organize our simple camp, lay in firewood, and have a leisurely swim and supper. It was a good island, with no company and a fine view of the sheer limestone cliffs that were now beginning to glow with first light, and if there had been any mosquitoes during the night I hadn't noticed them. I rolled out of the sleeping bag, stretched, pulled on the sand-gritty old levis, and shadowboxed my way over to a large maple that Tar and I each anointed in our morning micturitions.

The driftwood campfire was a bed of dead white ash without a spark of life; I built a tepee of twigs and shavings, adding more twigs as the tinder caught and staying with it and using larger sticks to assure a good bed of breakfast coals, humming some good-time song or other and noting once more the excellence of a river dawn flavored with the smoke of a cooking fire.

A hundred yards downstream the little island ended. There on the channel side the River ran beside a low bank that was strongly reinforced with a matrix of tree roots, cutting deeply under the roots and then deflecting, leaving a sandy point that shelved abruptly into deeper water off the point of the island. Just at sundown I had planted four "ditty poles" around that point, thrusting short poles of willow deeply into the solid mud of the bank, their baits in slack water just inside the shear line of the current. Two of the lines each had a treble hook baited with a ball of commercial catfish bait—a foul, glutinous dough based on something that had once been cheese, but which now transcended any Limburger ever created. It stank magnificently. Each of the other two lines was armed with a single hook that was baited with live "softshell" crawfish that Tar and I had captured the day before in a clear, rockstrewn feeder creek. I saw the cheese-baited poles first; they were lifeless, their lines hanging slack and baits leaching harmlessly into the clear brown shallows. It was another story with the crawdad poles. The first one was simply gone, pulled free even though I knew it had been thrust securely into the muddy shelf. The other pole was almost as motionless as the cheese-bait-poles had been, but the line was not sagging slack; it appeared taut, with something of a strain on the willow rod, and as I seized the pole and pulled it from the mud I felt the lively surge at the other end. It was a fish perhaps seventeen inches long, still "green," as they say, still with fight in it. The khaki-gray back blended down into spotted, silvery-gray sides, a slim, solid fish with fine running lines that ended in a deeply forked tail. Its scaleless cylindrical body had a slightly flattened head with fleshy "whiskers" on each side of the mouth. My fortune was made. I was a breakfast millionaire. Prize of my boyhood and the abiding joy of my later years—King Cat, the lordly channel catfish. I retrieved the other lines, whistled Tar out of the river, and headed for camp, getting hungrier just thinking about it.

With the fire almost burned down, I planted the blackened grill over the driftwood coals and put on the coffee and a frying pan filled with bacon. I skinned the catfish, making the cut behind the head and then slicing deeply into the body on each side of the dorsal fin, pulling it away with the catfish pliers that are mandatory river equipment. Then I stripped off the skin in one piece, snapping the head down and away, the guts going with it. A slit up the belly, the body cavity wiped out, and voila! Breakfast.

The bacon was done; I fed Tar half of it just to keep him civilized and ate the rest with my fingers. The last of the pancake flour went on the catfish; I freshened the fire with a few sticks of wood, and when the frying pan was sputtering properly the floured catfish was gently laid in the hot bacon grease. Bacon, of course, must be cooked slowly with due reverence; catfish should be fried quickly in grease that's almost, but not quite, too hot for the job, and it must be done with total concentration, taking care not to overdo the smaller parts of the fish.

Then it was done to a turn. I put the golden fish on the inverted lid of the frying pan and sat on the sand cross-legged with my millionaire's breakfast, facing the stone cliffs that were now flushed with the first of the morning sun. The succulent white flesh lifted easily from the fish's bones, flaking away sweet and delicate. If there is a tastier fish than a fresh-caught channel cat of that size, I reflected, what could it be? Fresh walleyed pike from a northern lake, or native trout taken from a high-country creek, each as cold to the touch as icicles and cooked within the hour of catching? Or pompano fresh-caught by a Tarpon Springs jetty and broiled Greek-style with paprika and olive oil? Consider halibut only a half-hour out of one of the Kenai Fjords of southern Alaska, charcoal-broiled and basted with lemon butter, or Atlantic salmon as they bake it in that country inn up the coast from Halifax. I've eaten them all—filleted, broiled, smoked, chowdered, and even baked Indian-style on a canoe paddle—and none could beat this Mississippi River catfish in anything but snob appeal. I thought of a certain Captain Marryat of the British navy who once honored western America with his presence. He didn't think much of the country in general and the Mississippi in particular, loftily declaring that it "contains the coarsest and most uneatable fish such as catfish and such genus," and Mark Twain reflected that "the catfish is a plenty good enough fish for anybody."

There was only one thing that bothered me as I sat there with my breakfast, watching the early sun glowing on the cliffs across the River. What had stolen that ditty pole so solidly fixed in the firm, half-dried mud at the end of the island? What manner of fish had come up out of the deep eddy in the night, prowling the shallows that might scarcely cover its back, finding a live crawfish so cunningly hooked through the tail, then seizing the bait and perhaps feeling the hook, and surging off into deep water, taking pole and line? Maybe it was just as well. Maybe, if that line had held, the great fish might have towed the whole damned island away, like something out of Dr. Seuss. It is the province of Mississippi River fishermen to reflect in such fashion. Next time, perhaps, we would lash a towboat hawser to one of the big maples with Tar tied to a hay hook for bait, and maybe take a Nantucket sleigh ride, island and all. How about that, dog? *Thumpthump* said the tail. Rave on, boss, and keep the bacon coming.

With the second cup of coffee I considered the point a bit more objectively. It could have been any of several different kinds of fish, and although none was quite capable of uprooting a river island there was no denying the fact that the River held fish that were mighty impressive animals. There is an axiom that big waters tend to hold the biggest fish—an obvious generality that may have some flaws but which has never been seriously doubted by travelers on the Mississippi. When Marquette and Joliet entered the Upper River in early summer, 1673, they met "monstrous fish" from time to time. One of those giants struck their canoe so violently that Père Marquette feared it might smash the frail craft, thinking at first that it was a large log or tree floating just under the surface. It could have been any of several kinds of fish, each of which might damage a bark canoe. You can expect almost anything of water as big as the Mississippi. Years ago, just upstream from Saint Louis and only a few miles from my back door, commercial fisherman Dudge Collins netted an eighty-four-pound bull shark, a species able to manage in fresh water for extended periods. It commonly occurs in Lake Nicaragua, where it has been known to attack swimmers, and while it is not one generally regarded as a man-eater, it is one of those sharks that should have something on its conscience.

"Bull shark, Tar! Think of that. An honest Injun, five-foot shark, hunting 'way up into our river. And you know what it was after? I'll tell you what it was after: fat, lazy Labrador retrievers that get on sleeping bags, and eat too much bacon, and track mud into the boat and then shake water all over the boss! *That's* what sharks like best!"

Thumpthumpthump said the black otter tail.

If the axiom of big water–big fish is valid, and there should be little doubt that it is, there should be a corollary that big ancient waters tend to hold big ancient kinds of fish. In any case, both axiom and corollary apply to the big and ancient waters of the Upper Mississippi, where fish of prodigious size swim unchanged from ten million years past.

Biggest of the bunch was the lake sturgeon or "rubbernose sturgeon," a primitive giant that has reportedly reached eight feet in length and three hundred ten pounds. This was very probably the kind of brute that struck Père Marquette's canoe that time— a collision that has been attributed to a very large catfish but which probably involved a big sturgeon, which is more likely to loaf near the surface.

The only lake sturgeon I've known personally was Old Oscar, who was in part-time show biz. Through most of the year he languished in a fish hatchery where he lay on the bottom of one of the ponds and subsisted on the leavings of hatchery fish. It was said that he had been a state employee for nearly half a century and that for over thirty years he had made an annual—and highly involuntary—migration to the Iowa State Fair where he had star billing at the state game-and-fish exhibit building. Each August he would be netted, loaded into a hatchery truck, and taken to his aquarium tank at the fairgrounds. He wasn't the biggest lake sturgeon extant. Years of confinement may have stunted his growth and he wouldn't go much over a hundred pounds. But speaking from experience as one of the battered, watersoaked flunkies who helped transfer him from hatchery truck to aquarium, a hundred-odd pounds of indignant lake sturgeon is enough. Once in the tank he always settled down and never did much, but a fish that size doesn't have to. He could just lie there on the bottom and be a hit.

My boss used to claim that two generations of fairgoers, and as many as four hundred thousand people each year, had been awestruck by Old Oscar. But when Old Oscar was finally gathered to the bosom of his fathers, the passing was unmarked by any publicity whatever, and lo! at the next state fair there was a new Old Oscar in the front tank. This rupture of continuity was never announced, and inspired me to ask the boss just how many Old Oscars there had been over the years. He gave me a long, cool look and replied: "The only thing you need to know is that the show must go on. Get out there and thump the tub."

Whichever Old Oscar it may have been, there could be no gainsaying his (or more likely, her) venerability. A lake sturgeon of a hundred pounds is an old fish as fish go; they don't even spawn until they are twenty years old and about four feet long, and then they're only beginning. They've been known to live 152 years, and have been said to top out at a length of eight feet and a weight of 310 pounds. They are surely the biggest fish in the Upper Mississippi—or at least they were. There can't be many of the old giants left in the River itself. They lay huge numbers of eggs once they reach maturity but may spawn only once every three years. Then, too, a lake sturgeon must overcome an infinite array of mortality factors to survive for twenty years before spawning. In the earliest days of commercial fishing, when lake sturgeon had no particular value as a commercial species and were often caught by nets in the Upper Mississippi and some northern lakes, they were discarded as useless. Millions of pounds of prime flesh and roe were left to the gulls, but this began to change shortly after the Civil War. In the 1870s there was a commercial fishing outfit at Sandusky, Ohio, that found a ready market for smoked lake sturgeon and caviar and was handling up to eighteen thousand Lake Erie sturgeons each year. In 1880 the catch of sturgeon in the Great Lakes was over seven million pounds, but had fallen to less than one hundred thousand pounds by 1917. A similar collapse took place in the sturgeon fishery of the Upper Mississippi, where the catch of lake sturgeons dropped over 50 percent from 1894 to 1899. The virtual disappearance of the lake sturgeon isn't surprising; as she approaches the first spawning of her life a female sturgeon may contain ten pounds of roe, making her a desirable catch at the very beginning of her sexual maturity. In 1909 a pound of sturgeon roe brought $1.50; one six-foot fish caught in 1911 yielded thirty pounds of caviar and the roe and flesh of that sturgeon netted the fisherman

$54.80 in a time when a man might not earn that much in two months of hard labor. Even the swim bladders of the big sturgeons were valuable. Used for production of isinglass, they were worth $1.00 per pound at the turn of the century.

Slow growth, late maturity, triennial spawning and a high price tag are enough to slate any commercial species as a born loser. And if those things aren't enough, plug in the needs for relatively clean waters and silt-free bottoms, with freedom to move into small feeder streams with shallow, gravelly riffles. It all adds up to a fish on its way out; after surviving as a virtually unchanged animal for millions of years, the great rock sturgeon finally came up against a set of conditions it could not survive for long. Today, if a canoe

in the Upper Mississippi is suddenly struck by an underwater object that might be taken for a floating log—it's a floating log.

The much smaller shovelnose sturgeon, or "hackleback," is far more abundant than its giant cousin and is still taken commercially in the Upper River. I've never seen one that went much over three or four pounds but those are valuable pounds. The other day I paid four dollars for a chunk of smoked hackleback that I could have put into my hip pocket, and didn't begrudge the cost. Even when shovelnose sturgeon costs $1.75 per pound, the fish markets can sell all they can get.

Like the lake sturgeon, the much smaller hackleback is a bottom feeder that can protrude its tubelike mouth to suck up crayfish,

insect larvae, fish eggs, and other animal materials from the riverbed. Yet, also like the lake sturgeon, it has been taken on hook
and line. It is far more abundant than the lake sturgeon and is the
only midwestern sturgeon of any commercial importance, but that
is all relative, for it is an uncommon fish at best. Shovelnose sturgeon are fish of strong water, usually found in the open channels
of big rivers in the main thread of the current, unlike the big lake
sturgeon, which is as likely to be found in large lakes as in big
rivers.

These two fish share an incredibly ancient lineage; there is
something antediluvian about a sturgeon of any kind—its vestigial
armor of bony plates so unlike the scales of higher fish, the cartilaginous skeleton with its rodlike notochord in lieu of a bonafide
spinal column, the sharklike tail and head, and that vague aura of
antiquity that seems to shroud creatures of this kind—animals that
have managed to hang on for millions of years, seeming to keep
pace with species that have rapidly evolved during the same time
span, but without making the same evolutionary progress. The
sturgeons carried certain survival equipment with them, to be sure,
and it was sufficient to ensure continuing existence if not steady
adaptation and improvement. But they have remained generally
stationary in the great parade of species—still in the parade, but

only marking time while the other species have gone on ahead. The great sturgeons will be dropping ever farther behind, back into the rear ranks, and finally out of the parade altogether and the River will flow without them.

There is another primitive giant that keeps company with the sturgeons. The paddlefish, or spoonbill, is a grotesque, scaleless, virtually boneless fish that looks something like a shark whose snout has been extended into a long, flat, paddle-shaped projection. It is one of a kind in the New World, found only in the Mississippi drainage system, and its only other relative exists in the Yangtze River system of China.

A strange creature, like no other. Remove that flat, fleshy paddle that constitutes the snout and there's a superficial resemblance to the sharks with the inferior mouth and sharklike tail. But there's nothing sharklike about the long, pointed gill covers like Mr. Spock's ears, extending far back beyond the gill slits. Those gill covers, as well as parts of the paddle and the fish's head, are dotted with tiny pits or pores that are highly sensitive sensory terminations that receive the countless stimuli such a fish must assess in order to feed, breed, travel, and survive.

In murky water—the prevailing condition of most of its home rivers—the paddlefish is usually a leaden gray, although in clearer rivers and lakes the fish may be a dark-bluish gray that is almost black. (This seems the case with most fish, an inverse relationship between darkness of water and darkness of fish. Smallmouth bass in clear lakes, for instance, are much darker than their brethren in somewhat turbid rivers, and the most extreme example of all is the entirely white catfish found in the waters of entirely dark caves.)

There is little doubt that the prime function of that ungainly paddle is feeding; as the paddlefish swims just over the bottom it sweeps the paddle back and forth, creating currents that stir the silts and debris of the river bottom and put them into suspension so that the paddlefish can take them into its gaping maw and pass the stuff over its fine gill rakers—a selective filtering system that retains the edibles and discharges the rest out the gill slits. Copepods, small leeches, worms, insect naiads, and all manner of small crustaceans are strained out of the muddy water and gulped down.

Taken individually, such organisms are unimpressive foodstuffs, but in the aggregate they constitute the greatest biomass in the Mississippi. They are the base of the River's animal food pyramid, and it is not surprising that the largest fish in the Upper Mississippi get right to the bottom of things instead of messing around with any intermediates. The largest water dwellers in the world—the baleen whales and the basking and whale sharks—also exist largely on the smallest foods available to them.

The maximum size of the paddlefish is a somewhat hazy and tenuous figure. The "modern" record is 143 1/2 pounds from the Upper Missouri, but an undocumented "200 pounds" often appears in the general literature. Dr. Karl Lagler notes that "the fish may exceed six feet in length and 150 pounds in weight." And although the average weight today is far less than that (a 30-pound paddlefish taken in a net is considered a big one) there is no doubt that the fish has a massive potential. My old friend Jerry Jauron, ex-conservation officer and indefatigable promoter of the Missouri River, once told me of two western Iowa youngsters finding a huge paddlefish trapped in a shallow Missouri River cutoff by receding waters, and how the two kids rodeoed that fish most of one morning, trying to ride it but unable to stick to the slippery, scaleless body of the big fish.

Years ago, in the course of writing a feature story for my newspaper, the legwork took me to Rock Island, Illinois, and the office of the United States Army Corps of Engineers. There I met a civilian whose job included diving around the big channel dams and their locks to effect minor repairs and check on the condition of the structures. He told me that he was working around one of the big lock bays of Lock and Dam 26 when a fish swam out of the bay and collided with him. It was, he said, a bit like being rammed by a sizable log that had been planed down to a flat point. It jolted him in spite of the fact that he was wearing a full hard-hat diving suit with weighted shoes and a work belt of tools. He never saw the fish, of course, but he sensed its living mass as it rammed him and then brushed past him, and he believed it was an extremely large paddlefish.

Did he often contact very large fish while diving around the big channel dams? He did, especially just below the toes of the dams where mighty turbulence plucks and tears at the river bottom, sometimes digging huge scour holes that the corps endeavors to stabilize with bargeloads of great limestone boulders. It may be necessary for divers to go down into those dark depths and examine the footings of the dams as best they can, working in near-blackness and pounding turbulence. One of the bad ones was Dam Number 10 at Guttenberg, Iowa, where the scour hole was said to be eighty feet deep and undercutting the dam. And while I've never been able to verify the rumors, I keep hearing of divers coming up out of such holes swearing never to dive there again—not while "there were such fish down there, fish as big as a man" that might endanger lifelines and air hoses. I suppose the story is apocryphal, as stories of that kind usually are. Still, I first heard it in the old fish-and-game quarters building at Sabula, Iowa, with Don Edlen or Officer Maury Jensen or somebody telling it while a wild night wind slammed at the windows, and it wasn't all that hard to give it credence, especially if you half-closed your eyes and imagined eighty feet of black, turbulent water churning above you, and things unseen but alive moving about you—heavy, sullen creatures stirred by your presence, brushing past you in the pounding alien darkness. Some men would surely balk at going down into such a world—and here's one of them. Whether any actually have or not is as much as the story's worth. But there is one thing I believe without reservation: If there is anywhere in the River today where giants could still exist, it would be in the highly oxygenated, food-rich depths of the great scour holes where no nets or hooks or barge traffic could ever reach them.

There are still paddlefish in the Upper River, although they are neither so numerous nor so large as they once were. There are probably fewer in the upper reaches than ever, but whether it is siltation of their spawning beds, pollutions of various kinds, breaks of some sort in food chain or life cycle, or just rampant overfishing, who can say? They are not especially common farther down near Saint Louis, but they are taken in fair numbers in spring trammel nets and constitute a rather steady item in many of the fish markets as "boneless cat." The name is taken from a slight resemblance to the catfish clan (no relation) and from the fact that the skeleton of the paddlefish is largely cartilaginous with just about enough ossification present to promote the fish into a higher class, *Osteich-*

thyes, as one of the bony fishes. But althogh it may have left the sharks, lampreys, and hagfishes behind, the paddlefish has kept many of its ancient characters, such as the notochord, the cartilaginous rod running the length of the body and originally serving as the body's axial support. This notochord is entirely lacking in the adults of the higher bony fish, but persists in the paddlefish.

My old commercial fishing partner and associate in a thousand fine adventures, Dr. Ed Kozicky, came in from the River one day with a story of a giant paddlefish. He and Curt Oulson had taken the big spoonbill that morning in a trammel net, had dressed their catch, and Doc had kept the fish's notochord to prove it.

Now, I've never removed the notochord from a paddlefish in my life, but there was no doubting the authenticity of the thing. Pale-colored, rather translucent, and resembling a rod of undercooked pasta, the notochord was about three feet long. "All right," said my old friend. "You're a learned fisheries type. Tell me, how long was the paddlefish?" I rose to the bait. I always do. I never learn.

Figuring that a notochord of that length would probably end in the caudal peduncle some eight or ten inches from the fork of the tail and extend to within not more than six or eight inches of the base of the paddle, I came up with a rough estimate of about five feet total length. Which would be about what—a ninety-pound fish? A *helluva* fish! After a while Doc stopped laughing, got hold of himself, and revealed that the notochord had been pulled out of a smallish paddlefish but had stretched almost twice its length in the process and had retained its stretched form. You learn something new about notochords almost every day.

Of the several families of Ancients in the Upper Mississippi, the sturgeons and paddlefish are the least voracious. Their smaller contemporaries, the gars and bowfins, are among the most savage piscivores in the River.

The bowfin or dogfish, *Amia calva*, is known locally downstream as "John A. Grindle," although I've never met anyone who could explain why. Whoever the original John Grindle was, he must have been a ring-tailed, case-hardened, center-fire son of a bitch with all the bark on. Probably some old commercial fisherman. His name and reputation have been fixed in natural history—denoting a fish whose line is almost as ancient as that of the sharks.

This bowfin, or dogfish, or grindle, isn't a big fish; I've never heard of one going over twenty pounds, and the average adult is probably only one-fourth that. The fish is a hangover from ancient times, and like any hangover it is not pleasant to behold. The body is a heavy cylinder of muscle armored with tough cycloid scales, with a blunt, slightly flattened head split by a mouth studded with sharp, strong teeth. A long, low dorsal fin extends down the length of the fish's back and seems to have a will of its own, undulating and rippling, and capable of slowly moving the fish through water without any other body movements.

As close as John A. Grindle ever comes to being handsome is in the spring when he wears his courting colors. His lower fins become a vivid paint green, and a dark spot, or ocellus, on his tail is bordered with brilliant yellow-orange. The long dorsal fin is dark green with a narrow olive band along the upper margin and near the base. The rest of the time he's just a dull olive above, with a conventional cream-colored belly.

I first met John Grindle while crappie fishing near a weedbed in Gun Barrel Slough, when he grabbed my bait minnow and took off for Memphis. He fought deeply with strong surges, and even when he finally ran out of gas and was brought to the boat he was still trying to bite through the hook. The bowfin fights like a game fish, but unlike some big bass and pike he shows few resources. He simply hauls away with brute strength. In the dead, heavy heat of high summer a bowfin doesn't fight hard or long; there's usually just one strong flurry and he gives up. At that time of year he may be somnolent and flabby, but in cooler water it's another story. Then the bowfin's muscles grow solid and when hooked he shows lasting power. In spring or fall a big dogfish can put up a vicious, prolonged battle that is the equal of any northern pike's the same size.

Old John doesn't give up the ghost easily, and rivermen tell of bowfins living all night on a riverbank. The fish's spongy swim bladder can function as a sort of primitive lung, and the bowfin can gulp air and "breathe" after a fashion, much as do the lungfishes. In cool weather and in the shade, a "grinnel" can live for twenty-four hours out of water.

They are voracious brutes that will attack almost anything in the water that might appear reasonable to attack. They will strike any natural or artificial bait that moves, especially in spring when the males are guarding egg-filled nests in shallow weed beds or watching over the newly hatched bowfin larvae that cement their snouts to plant rootlets about the nest. Even after the tiny bowfins are free-swimming fry, their male parent continues to guard and herd them, slashing out blindly at any real or imagined threat. Few bowfins taken on hook and line weigh over eight pounds or so, and the real lunkers are usually taken in nets by commercial fishermen who tell of giants running to nearly twenty pounds. Such fish are often ones that have been marooned in overflow ponds or oxbow lakes where, in such waters and cut off from the main channel, they may be netted by fishermen who will be able to catch nothing else. All other fish have been eaten.

Paddlefish and sturgeons are, as fish go, degenerates; whatever they were to begin with, they have lost most of their scales and become long-snouted grubbers with feeble jaws. The bowfins and gars, on the other hand, are Ancients that have not really degenerated but have simply failed to make any real evolutionary progress since the age of dinosaurs. They are the only survivors of the ancient ray-finned fish, the actinopterygians, that were the dominant fish of the Mesozoic era. The rayfins were the evolutionary step just before modern fish, the teleosts, and almost all the rayfins migrated to salt water where they continued to evolve. The gars and bowfins remained in fresh water—and never changed.

Of the four Ancients found in the Upper River today, the bowfin looks the most like a modern fish with its flexible, rounded, overlapping scales. The paddlefish, of course, has practically no scales at all, and the sturgeons have only several rows of heavy, conical scales sometimes called "bucklers." The gars are armored with strong, bonelike, diamond-shaped ganoid scales. Joined edge to edge, rather than overlapping shingle-fashion as with most fish, these scales have cost the gars much flexibility. But while they may not bend as easily as other fish, they're a whole lot tougher. Spearing a gar with an ordinary fish spear is like trying to gig a brick.

Indians once fashioned arrow points from gar scales, and old-time southern farmers sheathed the wooden moldboards of their plows with gar skin.

The biggest of the American gars, the alligator gar, is a behemoth of the Lower River, rarely found upstream from Saint Louis. It rivals the lake sturgeon as the largest fish of the Mississippi drainage; there are reports of alligator gar that measured nearly ten feet long and weighed three hundred pounds. Technically, it may be a fish of the Upper Mississippi—but not by much. If it does get past the mouth of the Missouri River, it's only for a few miles.

The gars of the Upper Mississippi are far smaller fish, but no less tough and voracious. The shortnose gar, or billy gar, has a tooth-filled snout that makes up about half the fish's total head length. The longnose gar's snout is much longer and narrower, about twenty times as long as it is wide. This is the larger of the two fish; a longnose gar may be several feet long and weigh as much as twenty pounds. The much smaller billy gar has to hurry to make three pounds.

These are cylindrical, cigar-shaped fish, long and slender, with bony snouts studded with sharp teeth. Like the bowfin, their air bladders are connected with their throats and are highly vascular. We see them in isolated oxbow lakes in the hottest, heaviest days of summer, thriving in "dead" waters that hold practically no oxygen. They occasionally surface to gulp air, opening and closing their toothy beaks with an audible snap. This is called "breaking," and if it happens under your elbow in a quiet backwater where you're drifting half asleep in a canoe, your startled reflex can produce dramatic results—especially in a canoe with no keel.

Gar are deadly fish-hunters. They are not as likely to chase food fish as hunt them, stalking their prey by moving slowly toward it by means of rapidly vibrating fins but with no apparent body movement. Looking like nothing more than a waterlogged stick drifting through the water, the gar closes on its victim and makes a swift deadly lunge that is almost too quick to follow. I have watched them do this in clear shallows where they have stalked my floating lure, seemingly indifferent to it until the bait was within striking distance. Far more often than not, they hit baits and lures with

impunity; it is almost impossible to set a hook in that bony snout. The only workable gar lure I've ever used was a six-inch length of quarter-inch braided nylon cord that was carefully unraveled to form a hank of fine nylon fluff. This is simply tied to the end of the line without a hook, and the toothy snout of a striking longnose gar is inextricably tangled in the stuff. People along the River do eat gars now and then, and the price has recently been as high as twenty-five cents a pound. I've never eaten any and don't figure I've missed a whole lot. And I wouldn't touch a gar's caviar on a bet—not after knowing Neil Nelsen.

Neil was one of the stalwarts of the old Iowa rough-fish-removal crew, a raffish bunch that was tireless in its efforts to purify the state's lakes. With seines up to a mile long, working in open water in summer and under thick ice in winter (and occasionally losing a truck in that latter circumstance), they would net vast quantities of carp, buffalo, gar, and other rough fish. The carp and buffalo were sold; the rest were given to farmers for fertilizer or simply buried somewhere.

While Neil was doing this, his wife Beulah ran their little restaurant and tavern at Orleans, a small settlement on the narrow spit of land lying between East Okobojii and Spirit lakes in extreme northwestern Iowa. The place was a favored hangout for us fish and game types, and the Nelsens were well regarded by all except the guy whose summer cabin was next to the tavern. The wholesome lakeside atmosphere, he felt, was tainted by the uncouth tavern next door, and although any of us would agree that Beulah's place might not be the classiest joint in the world, it wasn't that bad. Besides, that neighbor's property was no great asset to the community, either; during the summer he kept a small flock of chickens in a pen that was closer to the Nelsens' windows than to his own—a situation that Beulah had pointed out on several occasions. Neil pondered the problem. The Viking mind grinds slow but exceeding fine, and opportunity knocked one day when the neighbor happened to complain that his hens weren't laying.

"What are you feeding?" Neil asked.

The neighbor was feeding laying mash and cracked corn.

"You tried any fresh fish eggs?" Neil wanted to know. Of course he hadn't. What good were fish eggs?

"I'll show you. I'll get you some," said Neil.

It was spring, and the rough-fish crew's next job produced a number of female longnose gar that contributed their eggs to the Nelsen cause. Neil dutifully delivered the fresh gar eggs to the neighbor, who promptly fed the chickens.

Next day the man came raging into the tavern where Neil and Beulah were having coffee with a couple of local walleye fishermen. "Damn you, Nelson!" the man shouted. "Come over to this window and look at my chickens!" Neil obeyed, and surveyed the next-door scene with Nordic calm. The yard was strewn with very dead chickens.

"They don't look too good," Neil offered, sipping his coffee.

"Don't look too good?" the man bellowed. "Of course they don't look too good! Godammit, they're *dead!* It was those fish eggs. They were poison!"

"Well," Neil replied thoughtfully. "You said you wanted them to lay. Look at 'em. They're laying all over the pen."

It wasn't the first time Neil had confused active and passive verbs, but it was the last time he ever had any chickens next door.

More of the Ancients range together in the Mississippi River system than anywhere else in the world. Nowhere else are paddlefish, several kinds of sturgeons and gars, and the bowfin found in the same waters. They are curios, relics of an age so incredibly remote that we can scarcely conceive it, predating even the earliest mammals. The paddlefish and sturgeons are probably on their way out. Time is not their enemy—they have proven as well as any vertebrate that they can endure great doses of time. But when time is allied with deteriorating habitats, their endurance wanes and their midnight draws closer. And although the gars and bowfin appear to be surviving and even flourishing, there are signs of weakening. The great alligator gar is far more rare than it was—partly because of overfishing (the huge fish are actually taken on hook and line, with big hunks of meat for bait and a rifle or pistol as final authority) and because the productivity of their favorite river, the White River in Arkansas, has fallen off greatly in recent times. No one is sure why, but it is thought that the big upstream impoundments on the White have cooled the river below its optimum spawning temperature for the 'gator gar.

The other gars, and the bowfin, seem able to endure the worst environmental insults we can throw at them, and are apparently doing as well as any fish can be expected to. But the grim fact is that the 'gator gars are anachronisms that have greatly outlived their time. Maybe they will continue to make it as the cockroach and turtle have, their evolutionary progress apparently checked in a world where change is constant. They may even outlast us. But sooner or later their inability to evolve will be the death of them, for some unendurable change is certain to occur. Until then they are living on borrowed time in a world teeming with more advanced fish—the teleosts, the bony fish with up-to-date skeletal structures and modern ways of doing things. The Ancients are coexisting and competing with some of their own lineal descendants, a curious situation analogous to one of us sharing an elevator with an *Australopithecus*.

By today's measure, one might think that the Upper Mississippi was a teeming aquatic wonderland a century or more ago; riparian industry was relatively light, upland farming hadn't begun to pour its immense silt loads and chemicals down the watersheds, and the heyday of heavy commercial fishing was yet to come.

But enough inroads had been made into the pristine Upper Mississippi fishery to begin causing some alarm shortly after the

Civil War. A United States commissioner of fish and fisheries was appointed in 1871 to investigate the problems confronting American food fishes, and in 1872 the office's duties were broadened to include fish propagation.

In the Upper Mississippi, the commissioner's first effort was the stocking of twenty-five thousand American shad just above Saint Paul. A valuable food herring, this shad is a marine species that migrates into fresh water for spawning, and although successful establishment in the Upper River would have meant a long annual migration to the Gulf of Mexico this was rationalized by the fact that shad in China made a longer journey up the Yangtze. Even in 1872 shad-stocking was admittedly a long shot, but success would be of "such transcendant importance to the food supply of the country, and the cost of the experiment so trifling, that it would be inexcusable not to attempt it."

From 1874 to 1884 nearly one and a half million American shad were stocked in the Mississippi. There are some cloudy reports of a few being caught as a result of all this, but by and large the stocked fish seem to have evaporated, as stocked exotics often do. So did the Atlantic salmon that were stocked in the Upper River— apparently on the premise that the River was suitable salmon habitat because it generally lacked dams or other major impediments to migration. But even as native, sea-run shad and salmon failed in the Upper River, a big minnow newly arrived from Europe became naturalized almost overnight and proceeded to bully its way into a position of overwhelming dominance.

The "German" carp, originally out of Asia, was a cherished food fish in the Old World where it thrived in muddy ponds and streams —a fecund, fast-growing fish that was tolerant of a wide range of marginal water conditions and was considered highly edible in the bargain.

The carp may have been introduced by a Mr. R. Poppe in 1872, and by the United States Bureau of Fisheries five years later. There was apparently no preliminary study made of the fish and its requirements, nor of any interactions with its habitat; the Fish Commission simply imported several hundred adult carp from Europe and began propagation.

Years ago, while prowling through an ancient issue of *Forest and Stream* magazine, I ran across a singular bit of news:

Washington, D.C., June 6, 1877. The introduction of carp in waters of the United States has engaged the attention of Prof. Baird, chief of the Commission of Fish and Fisheries, and he has already imported some of the best varieties of German carp, which are regarded as the best in the world.

Four hundred and fifty of these fish were recently received by the steamer *Necker*, and they have been placed in the ponds at the Druid Hill Park near Baltimore in charge of the Maryland Fish Commission. They will be kept for breeders and soon ponds will be constructed near Washington with the view of obtaining as many young fish as possible to stock in southern waters. The carp does not promise to be as valuable as the mackerel, shad or salmon, but is a fish of great commercial importance for the reason that many can be kept on a small body of water.

. . . It is a fish eminently suited for southern streams, and it is proposed to stock those streams with them . . . The meat of the carp is very palatable and as a food fish it stands high though, as stated above, does not equal mackerel or salmon.

This was the first large shipment of carp to America by any agency. Many shipments had already been attempted by private concerns and individuals, however, and carp probably existed in midwestern waters in the early 1870s.

In 1873 an Illinois fisheries worker, Dr. Sylvester Bartlett, received a letter reporting a carp taken in the Mississippi River near Quincy. The excited fisherman wanted to know if either Illinois or the United States Department of Fish and Fisheries was stocking carp in Illinois waters.

Dr. Bartlett wrote back: "As we value carp too highly to experiment with them by putting them into our rivers, it must have escaped from a pond or livebox. It nevertheless demonstrates the practicability of eventually stocking our streams with this wonderful fish."

The imported stock thrived under the tender, loving care being lavished on it, and there were soon enough fish to begin shipments. With much fanfare and political hurrahing, young carp were distributed from coast to coast. States anxiously awaited their quotas for stocking, and when a modest shipment of young carp arrived it might be greeted at the railway station by a brass band and paraded through town on its way to the river or pond. The first carp arriving in the upper Midwest were usually held in ponds, many of which were specially built for the purpose. It was somewhat later, when good supplies of this wonderful fish were assured, that carp were freely stocked in streams.

By the mid-1880s, stocked and feral carp were spreading widely in the Upper Mississippi and most of its feeder streams. The late Harriet Bell Carlander, in her superb monograph *History of Fish and Fishing in the Upper Mississippi River,* told of the first carp taken near Lansing, Iowa. It was in the late 1880s and the fisherman, one Sever Olson, didn't have the foggiest notion of what he had caught. Neither did any of the other commercial fishermen. Sever took his fourteen-inch mystery fish into town where it was quickly (and happily) identified by the local druggist, a German immigrant named "Doc" Nachtwey. We can assume "happily" because Nachtwey paid Sever a dollar for the fish—and later paid a dollar each for the three or four other carp taken that year in the River. But that was the last year he or anyone else could afford to

pay a dollar for every carp caught. The new fish lost its novelty
fast. By 1894 carp had been reported from Missouri, Illinois, Iowa,
and Wisconsin, and the total catch was nearly half a million pounds.
By 1899, only five years later, the commercial catch had increased
sixfold. The honeymoon was definitely over; in the 1909–1910 bien-
nial report of the Iowa fish and game warden there was a pointed
reference to "the despised carp," and it has been a gloves-off
donnybrook ever since.

As it turned out, the "wonderful fish" hadn't needed much
encouragement. It was a generalist, hardy, thrifty, and adaptable,
and made itself at home in the same manner as the Norway rat
and English sparrow. It is a fish of many admirable qualities. The
carp is a virtual art form in China and Japan. In Europe it is a
carefully cultivated food and game fish, and our revered Izaak
Walton, patron saint of anglers, wrote in his *Compleat Angler*,
"The carp is the queen of rivers: a stately, good and very subtle
fish. And my first direction is that if you fish for carp, you must
put on a very large measure of patience." Europeans covet the
carp for sport, stalking it with all the care given pooled salmon,
never handling bait with bare hands, and wearing soft-soled shoes
to avoid any vibration as the angler steals along the stream bank.
It is said that in England a carp weighing over twenty pounds is
an angling trophy that may be rushed to the nearest taxidermist,
and it is reported that the angler who caught England's record carp
"looked as if he had been in heaven and hell and had nothing more
to hope for from life."

Carp in America fight with the same dogged, powerful cunning
that Europeans admire; they are no less combative for having changed
their flag. And although there are American anglers who specialize
in big carp, and while a fish weighing twenty pounds or more can
absorb one's full attention for quite a while, I've yet to see the
lucky angler heading for a taxidermist. Just as I've never seen a
mounted carp, I've never seen a mounted starling, either. It is a
matter of respect—and for the carp, being a staunch and gamy
fighter on hook and line isn't enough to elevate him into exalted
ranks of game fish, or even up to the pan fish. A rough fish he
remains, by personality and disposition, and is likely to remain so

until he is the last survivor in our envenom'd waters—which is also likely.

Much of the carp's off-color reputation has developed, I suspect, from the overwhelming adaptiveness of the fish. It has a lot in common with the dandelion, another immigrant that is also ornamental, highly edible, hardy, and fecund. Both carp and dandelion would be treasured species if they just weren't so prolific and adaptable, and could be depended upon to more or less stay where we wanted them. As it is, they have simply moved in and taken over, and their familiarity has bred our contempt.

If the common carp carried, say, only a thousand eggs per pound of body weight, needed highly specialized conditions for spawning, and could thrive only in lakes with a mean temperature of 72 degrees and high oxygen content, it would be in relatively short supply and probably coveted as a rather classy game fish. As the carp decreased in numbers certain game fish and pan fish would surely show concurrent increases, lowering their relative status at the same time the carp's was being raised. Supply-side carpenomics.

The carp did not move into an environmental vacuum when it arrived here; all ecological niches were comfortably occupied by native fishes and there were no real vacancies even though certain fish populations were already being thinned by reckless overfishing. Rather, the carp wedged itself into existing environments whether the current occupants liked it or not—and more often than not, they didn't.

The carp is a messy eater, rooting for much of its food like a hog, tearing up the stream or pond bottom and muddying the water and being a general nuisance. For a long time it was believed they were largely vegetarians, tearing out the succulent shoots and rootstocks of aquatic plants and turning rich ponds and marshes into sterile sinks. In his 1904 report *The German Carp in the United States*, L. J. Cole stated that "the evidence seems to be pretty strong that in general they are very destructive, and are probably in part, at least, responsible for the great reduction of wild celery and wild rice that has been noted in many of our inland marshes in the last few years." Even two of the nation's premier fisheries scientists, Stephen Forbes and Robert Richardson, reported in 1920 that carp were omnivorous feeders whose staple foods were vegetable matter. To this day, most fishermen and even some fisheries workers believe the carp is a dedicated vegetarian that eats the odd snail or fish egg only by accident, and up until 1946 there were few scientific studies to prove otherwise.

For three years, until 1949, Iowa fisheries biologists collected a total of 687 carp in a variety of natural conditions and carefully examined their stomach contents. The results left no doubt that under normal conditions the diet of carp of all sizes is predominantly animal material—aquatic insect larvae, small crustaceans of various kinds, and snails. Larvae were most important, with midge larvae heading the list. There were usually some plant materials in the carp stomachs, to be sure. Dead plantstuff had almost surely been taken while feeding on animal matter; there was evidence of green plant parts being eaten by choice, but by only a few carp in any one collection. Somewhere, perhaps, the common carp subsists on salads. But in Iowa, at least in the course of that study, the carp eats animal protein.

It all added up to the same thing, though. When carp took over a pond, lake, marsh, or even a river, the water seemed to take on a sick murkiness, and many of the aquatic plants, both submerged and emergent, began to thin and then vanish.

Aquatic plants in balanced natural situations are the essence of good clear-water habitats. They are the lungs of lotic and lentic environments, their respiration charging the water with oxygen. They are protective cover for game-fish fry; they also host the myriad larvae of insects and amphibians, crustaceans, and other creatures near the bottom of the food chain, and are prime provender for waterfowl. Carp eradicate these essential aquatics in the course of their feeding and rooting for larvae and other animal foods; carp-infested shallows may be strewn with the floating bits and pieces of aquatic plants torn up and broken by feeding fish. Even worse, perhaps, such rooting muddies the water, blocks sunlight, and prevents photosynthesis. The plants vanish from the sick and muddied waters; oxygen depletion sets in as a result of decreased plant respiration and because muddied water tends to heat more readily in summer than does clear water, lessening the water's capacity to hold dissolved oxygen. (None of which makes much difference to the carp; it can exist at low oxygen levels at water temperatures up to 96 degrees). Sight-feeding fish may be hard hit. Not only is the carp competing directly with them for animal foods, but the food chain whose bottom link is connected to aquatic plants is broken, and the balance of occupancy swings strongly in favor of the invading carp. The changing habitat may not be entirely to the carp's liking, but it endures. As food becomes harder to find the rooting carp becomes more diligent in its ruinous efforts, locked into a cycle of poverty like the hardscratch farmer who passes his desperation and suffering on to his land.

Carp can be ruinous to a small duck marsh, which quickly becomes a featureless expanse of murky water with a notable lack of aquatic vegetation or anything else. If such a marsh has a water-control structure at the outlet, however, its terminal illness can be reversed almost overnight. Completely drained with all carp removed, the barren silt bed exposed to air and sunlight, the dry marsh literally explodes with greenery. Long-dormant seeds buried in the marsh bottom germinate, and such marsh-edge plants as cattail and bulrush begin coming on. The water-control structure is closed. The basin begins to refill with clear water. Floating and submerged aquatic plants return in strength. So do waterfowl, and muskrats and minks, blackbirds and marsh wrens, and the spring and summer sounds of frogs and toads. But sooner or later the persistent carp will almost surely return. Will it be as eggs, carried on the feet of wading birds? That's a common hypothesis. More likely, though, they will come back as the bait minnows of fishermen who don't know the difference between a common shiner and a fingerling carp, and who don't particularly care. Carp may even be deliberately stocked by inveterate carp anglers or certain soreheads who have a grudge against the people in Fish & Game. Anyway, the cycle starts all over again.

None of this is meant to infer that carp completely dominate their invaded habitats and exist there in wholly pure populations. Nowhere in the Mississippi are carp found to the exclusion of all other native species. But there is no doubt that native fish populations, especially the game fish or "fine" fish, suffer from the changes wrought by carp. And it is not just certain basic changes in the game fish habitat. There is only so much *lebensraum* in any body of water, and the space occupied by a big, brawny carp is that much less space to be occupied by something else. There are more subtle ramifications, too—the habitat changes caused by carp may be of some benefit to certain "rough" native species of fish, giving them an advantage that never existed in pristine populations.

And so the once-coveted carp has come to be generally reviled, denigrated, and regretted—another textbook example of man's interference with environments that he didn't really understand. The German carp, prized though it may be in Europe, has about the same social standing in America that the European rabbit has in Australia.

But no fish is all bad, least of all one that is as productive, fast-growing, and edible as the common carp. It is not even possible to say how bad this new kid in the neighborhood has been, because the presence of carp has probably masked a lot of environmental reverses that would have occurred even if the carp hadn't been brought over. It is comfortable to reflect on how clear our waters would be, and how brimful of splendid native game fish, if our grandfathers hadn't been so stupid about exotic introductions. But there were reasons for those introductions of non-native fish in the first place: the natives just weren't doing all that well. Even if carp hadn't come along, our native game fish didn't have the rosiest of futures. Carp were just another problem, and we will never have a real perspective of exactly how significant that problem has been.

In all fairness, neither carp nor their sponsors have been all wrong. The lowly carp has turned out to be the predominant commercial fish of the Upper Mississippi. In 1894 there were only 453,000 pounds of carp taken in the Upper River; in 1950 the carp catch was 5,824,000 pounds. The annual harvest continues to be about six million pounds. In 1894 the leading commercial fish was the buffalo fish, with nearly six million pounds taken, and by 1950 this had declined to less than a third of the former total. Again, was the carp a cause or effect? Did it "crowd out" the native buffalo-fish to some degree, or was the buffalo a casualty of changing river conditions and heavy overfishing? Probably all of these. But good or bad, fair or foul, the common carp today is the linchpin of fish populations in our larger warm-water rivers and lakes. The fish appears to be here to stay, barring some highly specific piscicide that can be made for a penny a pound, applied for a dollar a ton, kills only carp, and is biodegradable. And should that miracle ever come to pass, a second miracle would be needed to pacify the thousands of commercial fishermen, fishing-lake operators, and hook-and-liners who dote on the carp for money and sport.

The carp is a burly fish whose record weight is a little over 83 pounds. That was from a lake in South Africa. The angling record in this country is a 55-pound, 5-ounce fish caught in Clearwater Lake, Minnesota. I almost said "hook-and-line" record, which wouldn't exactly be right. That record, as far as I know (as well as the carp record for the Upper River) is held by the 59 1/2-pound carp taken by commercial fisherman Paul Slater of Fort Madison, Iowa, in 1955. The huge carp was an even four feet long and was caught on a small Number 1 trotline hook baited with a bit of cut bait no larger than a pea—a common practice among some rivermen who may get four trotline baits out of a single angleworm. As Paul was lifting the trotline he realized he had a very large fish of some kind, but the giant carp offered surprisingly little resistance. Paul eased it up to the side of his aluminum johnboat, slipped his big landing net over the fish's head (as far as it would go), put his other arm under the fish while tipping his boat far over, and with a single heave rolled the fish into the boat. *Then* the carp knew it was hooked. By the time Paul finally tamed that fish, "everything loose in the boat had ended up in the river." The fisheries boys read the annuli on that fish's scales and put its age at from thirteen to fifteen years.

Although the average weight of carp falls far short of that, the fish ranks today as one of the two largest scaled fish in the Upper Mississippi. Its native counterpart is the largest of the buffalofish, the bigmouth buffalo, another "rough" fish that is broadly similar to the carp in size, habitat, and body form. Like the carp, it usually grows largest in lakes and ponds, and an eighty-pound buffalo was once taken from Spirit Lake in northwestern Iowa. In the Upper River they commonly reach thirty pounds, which is about the heaviest that carp commonly run.

The two fish share the same habitats, size range, and general body form, but that is where it ends. Buffalofish are members of the sucker tribe, *Catostomidae,* while carp belong to the minnow family *Cyprinidae.* Carp are clad in scales of tarnished brass with yellow underparts; bigmouth buffalo are often bluish-green on their upper works with bellies of cream or white. A carp's head tends to be somewhat conical and a bit smaller in proportion to the rest of the body. The head of a bigmouth buffalo is big, with a smoothly elliptical shape that looks downright elegant on that big, robust body.

It is a powerful fish, solid and muscular. We were reflecting on this the other day while eating some deep-fried buffalo that Jack Downs had just created. Beneath a shade tree out at "Nilo Jack's" is as fine a place as you'd want to be on a summer Sunday, and the river yarning doesn't come any better than from friends like Ed Kozicky, "Paddy" Paddock and his son Rex, and Curt Oulson. The Paddocks and Curt live in that swingingest of all swinging river towns, Grafton, Illinois, at the junction of the Illinois and Mississippi rivers, and are sometime commercial fishing partners. We were talking buffalofish, and Rex had just said that he'd as soon eat prime bigmouth buffalo as anything that had fins, and that led into a discussion of *big* buffalo, and got Curt started.

"A farmer wanted me to net his pond up here a ways. No, John, I *ain't* gonna tell you where. I set a trammel net that was six feet deep and took five buffalo out of that pond that added up to 259 pounds. Over fifty pounds apiece, and like peas in a pod. And you know, there was another fish in there that we couldn't hold. It tore right through a net that had been holding fifty-pound fish. Rex, that was Number 6 net twine. What's that supposed to hold?"

"Runs to about twenty-pound test. Good strong netting."

"Well, it wasn't strong enough," said Curt.

I like buffalo. I like the feeling of catching the tail line of a buffalo net with the grab hook, and lifting the net until the tail line can be secured on the bolt that is fixed to the gunwale of the boat, and then heaving up the hoop net and feeling the considerable weight of it, and just as the top of the net breaks the surface of the river there's a mad surging flurry that wets your oilskins, and you can see the broad backs of large buffalofish—not the tarnished brass of carp, punctuated with black where the scales intersect, but that special bluish-green with flashings of cream and white, and those big, blunt, bullet heads of buffalo.

I've never caught a buffalo on hook and line. Hardly anyone does. Back in 1975 a lucky cuss named Dave Hulley took a 47-pound, 2-ounce buffalo out of an Indiana lake, but I've never heard of one even close to that size being caught on a hook in the Mississippi. None of the three species of buffalofish in the River—the bigmouth, smallmouth, and black—are likely to take the kinds of baits and lures offered by anglers. Their diet is largely small crustaceans of various kinds and they just cannot be conned by artificial lures and the usual baits, although they are sometimes taken on trotlines baited with very small dough balls.

At the right place and time, say, at the shallow inlet of a large lake during spring spawning, carp may even be taken on the fly rod with such artificials as maribou streamers. At almost any time during the open-water seasons carp can be caught on a wild variety of soft baits such as dough balls even though, technically, these fish are on a general diet of small larvae and crustaceans. The fact that they can be caught on certain baits and lures that do not resemble their usual foods is another clue to the generalist, adaptable character of the carp. In contrast, grown buffalofish go chugging along through the middle depths inhaling zooplankton and selfishly depriving us sports (and perhaps themselves, as well) of all kinds of fun.

My boyhood river was the south fork of the Skunk in central Iowa, a modest little tributary of the Mississippi that held boundless potential for Huck-Finning away the summers—and I exploited that potential to the hilt.

There was a certain hole just below a bend, a little pocket no more than ten feet wide and twenty long, created by the digging action of the current under a fallen sycamore that had caught a pile of driftwood. On a midsummer day when the sun was right, you could almost see the sandy bed of the river through seven feet of brown water. It was one of those places that naturally catch any river rambler's attention, boy or beast. If you crawled up to the edge of the bank quietly and very slowly, there was usually something to see: a big snapping turtle, maybe, or a mink, and almost always a sizable fish or two. And so it was, on a groundhog safari one long-ago June, that I made my customary approach through the woods, eased up to the edge of the bank, and saw the Great Fish.

It was almost fully revealed just upstream of the big log, hanging nearly motionless in the current, a dim but distinct hugeness in the golden water. I'm sure it wasn't as big as I remember it, but it was easily the biggest fish I had ever seen in my small empire. Dim as it was, I knew it was not a carp. There were no black fleckings, and the general color of the fish did not blend with the water color; instead, it seemed of a cooler, grayer tone that contrasted with the golden-brown of sand and water, with a great broadness of back and bluntness of head. I reckoned it to be at least as long as my arm; as I watched, the big shape began to fade, going deeper and retreating back under the log drift, and my knees were growing watery and I had to sit down. It was the most singular and notable event in my singular and notable fourteenth summer. Groundhogs forgotten, I took off for home and swapped rifle for fishing rod and tore back as fast as I could. The big fish was nowhere to be seen. Still back under the drift, which was good. I applied a slip sinker and a carefully hooked "red wiggler" and let the current take the bait back under the drift—and I began a vigil that went on until dark. Nothing. Not a tap. Nothing the next day, either. I carefully described the fish to two of my senior mentors, Obe Blair and Hans Shockley, and they concurred that it was a buffalofish of some kind; about the only thing it could be, if it wasn't a carp. They also concurred that I had about as much chance of catching that fish as I had of getting rich. They were right on all counts. I saw the big "gourdhead" (Hans's name for the fish) several times after that, so I knew it was still using the hole under the log drift. I tried crayfish, dough balls, several breeds of angleworms and even some catfish "stink bait," all to no avail. The huge buffalo refused to play my game and I never did find out what his was, and so we ended the summer in a standoff. I wish I could give you a more exciting close to the story than that, but there it is. I've almost caught a lot of big fish.

That buffalofish and the big carp that shared my home river had much in common: each species is well adapted to silty water, can take relatively high water temperatures, and has no fussy spawning needs. Both fish have high reproductive potentials and can heavily populate even so-so environments. Yet, the immigrant carp has the undoubted edge on the native buffalo and has come from behind to outnumber it in most places.

The various buffalofish are the giants in the tribe of river suckers that includes a host of river carp suckers, quillbacks, chub suckers, and redhorse. Aside from the buffalo, the redhorse are the only ones much used for food—and then only locally. Years ago, it is said, people in the Upper Mississippi valley netted spring runs of redhorse suckers for food and salted the fish for winter use. If this is still done at all, we haven't heard of it. Redhorse are not generally in fashion.

These redhorse suckers are never big, rarely over eighteen inches, and almost never taken on hooks. Their flesh has many small faggot bones. Withal, anything but game fish. But don't knock them—at least, not before you have spent part of a late October night on some Ozark gravel bar, fueling the fire beneath the big iron kettle of melted leaf lard and making sure it keeps smoking hot. From downriver comes the far sound of an outboard motor; there is a spark of bluish-white light in the distance, a gasoline lantern mounted behind a reflector in the bow of the johnboat. Then the "boys" arrive, beaching the boat and lifting out the box of redhorse suckers they have speared in the cold, crystal pools of the spring-fed river. The fish are quickly "sided," their icy flesh scored so the faggot bones will cook out, breaded in a special formula, and dropped into the hot lard. The golden sides come out of the kettle steaming hot in the cold air, sweet and delicate. Doubtless there is grub as good as that somewhere, with company as convivial, and doubtless I shall never find them.

All suckers are "coarse" fish, which is another way of saying they are "rough" fish, distinct from the more exalted "fine" or "game" fish. There is even a subdivision of the latter, the "pan" fish, which are those smaller game fish that fit nicely into a pan. Such arbitrary gauging is done on the broad assumption that "rough" fish are social inferiors with coarse flesh and manners to match, generally unhookable and inedible. Game fish, on the other hand, are those species that require good neighborhoods, can be readily taken on the hook (especially on artificial lures, and the smaller the lure, the better), struggle "gamely" with the fisherman (the gamest usually put on a flashy show of jumping), and are pure ambrosia when broiled with sliced almonds or whatever.

I'll subscribe to all that—with some reservations. But there is considerable overlap. Carp, for example, may not be fussy about water quality but they will take artificial lures at times and I have never caught a carp that couldn't, say, outfight a walleyed pike or a Dolly Varden trout—and I have caught sizable examples of all three fish. Further, a common carp can be extremely good to eat. The ancient wheeze about "planking carp"—carefully broiling a dressed carp on a plank and then discarding the fish and eating the plank—was devised by some wag who didn't know his fish. I've eaten carp broiled, deep-fried, smoked, and chowdered, and it was all good. In little rural cafés along the Upper River, the ubiquitous "fish sandwich" is likely to be either carp or buffalo.

Good as prime carp can be, prime bigmouth buffalo is even better by my taste. And although many river folk don't agree with me, freshwater drum (known variously as "sheepshead" or "white perch," depending on where you are) can also be fine eating. I have caught them on many kinds of live bait, and on small streamer flies, crank baits, and spoons while fishing for smallmouth bass in Mississippi tributaries. I always "side" or fillet these fish, and while cleaning them I have sometimes detected an odd odor never en-

countered in any other fish—a sort of fresh "salady" odor, somewhat lettucelike, unusual but not at all unpleasant. Although some writers have said the flesh tends to be fibrous and tough, I've not found it so. Nor can I fault the sheepshead's fighting qualities. Still, it is not a highly respected fish in the Upper Mississippi; one has to go farther downriver before the white perch gets the credit it deserves.

Although sheepshead are still abundant in all the pools of the Upper River (the annual commercial harvest is well over a million pounds), they have undoubtedly declined in numbers, for in the old days huge catches were sometimes taken with long seines from under the ice. In March 1884, a single seine haul made in the Iowa portion of the Mississippi produced 240,000 pounds of fish that were mostly sheepshead. In January 1898, a seine haul in the River just upstream from Dubuque took 28,000 pounds, including 600 pounds of walleyes, 9,000 pounds of buffalo, and over 18,000 pounds of sheepshead. And as the weight of the drum fishery has decreased, so has the weight of individuals. I don't think I have ever seen a freshwater drum that went over fifteen pounds, or hooked one that weighed more than five. We were apparently born a couple of hundred years too late to know the real lunkers; a century ago, hundred-pound sheepsheads were not uncommon, and old Indian kitchen middens have contained bones from freshwater drums that may have weighed up to two hundred pounds.

Thus, some of the "rough" fish, the Rodney Dangerfields of the Upper River—big, prosperous, and famous, but never gettin' no respect.

Game fish, as Al McClane defines them, are (in an angling sense) "any species of fish which can be taken by sporting methods and by reason of its size or vigor prolongs its resistance to capture." And although the Upper River has been sadly degraded as gamefish water, most of the sporting species are still abundant enough to comprise a highly active fishery.

Consider largemouth black bass, for example—the largest of our sunfish. The Upper River teems with largemouths. My good friend George Zebrun, field biologist of the Illinois Department of Conservation, has shocked and netted countless backwaters of the Mississippi in his state and is perpetually amazed at the number of largemouth bass that thrive there, apparently unknown and un-tapped by the average fisherman. "Our bass fishery isn't even scratched by sportfishing," he says flatly. "No, I don't know why. Too much cover? Too hard to fish? Or are the fishermen concen-trating on crappies, white bass, walleyes, or whatever? Beats me, chief. All I know is there are a hell of a lot of big bass out there that'll never be caught, or even see a bass lure."

The kind of places George has worked in for years are prime largemouth-bass habitat—those snag-filled, weedy, food-rich sloughs and backwater lakes of the Upper Mississippi. The largemouth is essentially a lake fish, seeming to do best in waters with little or no current, and has probably increased in the Upper Mississippi in the past sixty years or so, or, at least, increased in their ratio to other large game fish. They've always been there, of course, but a lot of good largemouth water hasn't been. The Upper River has become more lakish since the big channel dams, and this has had to be advantageous to the largemouth. Still, for one reason or another, this bass appears to be one of the better-kept secrets in the Upper Mississippi sport fishery. Relatively few anglers seem to be deliberate largemouth specialists—and the ones who strike a mother lode aren't likely to advertise it.

On the other hand, smallmouth black bass have almost certainly declined in numbers. The lordly smallmouth has some relatively fussy habitat requirements: it needs water of decent quality, which, in the case of the smallmouth, usually means flowing water of good oxygen content and moderate summer water temperatures, with reasonably low turbidity and pollution levels and beds of gravel or coarse sand substrate for spawning. In other words, prime small-mouth bass habitat begins to overlap trout habitat, and in their northern ranges it is not unusual for trout and smallmouths to exist in the same waters.

In its day, the Upper Mississippi was prime smallmouth-bass territory, which says a lot about the essential quality of the old River. The smallmouth is more a creature of moving waters than is its bigmouth cousin; one of the aliases of this fish is "river bass." But much of the River's original smallmouth-bass habitat was a casualty of progress—the alluvial gravel fans of tributary streams, stretches of rock-floored, rock-girt rapids, beds of boulders, and bright riffles were all smothered by silt or drift sand, and water quality in some areas began to degrade below the levels demanded by the lordly smallmouth. On the other hand, as some of the old natural habitats faded, man-made ones appeared. Eroding riverbanks, especially on the outsides of long bends, were being armored with "riprap"—great chunks of rock that covered the riverbanks for miles. Long before the channel dams, current-deflecting wing dams were built to protect banks or deepen the channel or both. Like the riprap, these were usually of limestone boulders and a fair approximation of natural river structures. Such masses of large rocks are wonderfully productive of the stuff cherished by smallmouth bass. Crayfish, caddis flies, hellgrammites, forage minnows, and other links in the aquatic food chain are at home, sweet home, in the endless mazes of cracks, crannies, and crevices that these jumbles of rocks provide. Working in the lower end of Lake Pepin with electro-fishing gear, Minnesota biologists have taken more smallmouth bass along the riprapped banks than in all other types of areas combined. That was at night, by the way. Riprapping may be banquet board for many game fish, but it gives them little or no shelter during bright, sunny days.

Speaking of Lake Pepin, in the early days it was evidently a game fish glory hole. There was nothing else like it, really—a huge natural impoundment of the main river by barrier sandbars at the mouth of the Chippewa. Unlike most conventional lakes, this flowing river-lake was an ecological mix of the best of the riverine and lacustrine, with a variety and abundance of fish rarely found in typical lakes. There were walleyed pike, northern pike, pickerel, muskellunge, smallmouth and largemouth black bass, sauger, yellow perch, crappies, bluegills, and rock bass, to say nothing of the catfish and sturgeons. Before the big channel dams blocked them forever, there were skipjacks that rose to the fly and tasted like shad.

Few of these Lake Pepin game fish were likely to ever grow as large as their cousins in the interior lakes. Big water the Mississippi might be, and home water for leviathans it might be, but in modern times at least it has not grown individual bass, pike, muskies, walleyes, or panfish as large as some grown in lakes. A five-pound smallmouth bass is a very big smallmouth from the Mississippi or any other river that I know of, but won't raise many eyebrows in the lake chains of northern Minnesota's border. The same thing is generally true of largemouth bass; I've seen some dandies taken from the Mississippi, but none to match the lunkers caught in farm ponds at the same latitude. In terms of sheer novelty and variety, however, few inland waters can compare to the Upper Mississippi. It has been said that it was once possible to catch twenty species of fish in one day from Lake Pepin. You would be hard put to do that now, but there are still countless places in the Upper River where you can drop a properly baited hook or offer the proper lure and expect just about anything to happen. From a fisherman's point of view the River can be a monumental crapshoot—with all the thrills and uncertainties that pertain thereto.

In the Upper River today, probably no scaled game fish is more fervently pursued than the walleye. Often called "walleyed pike," it is not a pike but one of the perch family. It embodies almost every plus a game fish can have. It is common but not *too* common, readily takes either bait or artificials, and is about as toothsome a victual as swims. No matter that purist anglers call it the "ring-necked pheasant of game fish" (for the same reason purist hunters call the pheasant the "cottontail rabbit of bird-shooting"), the walleye deserves its colors as a game species. No, Nick, it doesn't perform any flashy aerobatics when hooked, but it's a sturdy scrapper with nothing to be ashamed of.

Walleyes take their name from the opalescent quality of their eyes, an odd blind look that occurs in no other fish I know. Strange, large, milky eyes that glow in the dark like a cat's when light catches them, though what that has to do with the state of being walleyed is beyond me.

Frankie Heidelbauer and I were once fishing the opener for walleyes on Big Spirit Lake. Rather, we were trying to fish. All through opening day and the first night, and into the second day and the second night, a strong south wind of almost gale force kicked the big lake into a three-foot chop and made fishing nearly impossible. Not that wind and walleyes don't go together; a smart breeze holding steady out of one quarter for a day or so may concentrate forage fish along the lee shore of a lake and draw walleyes to within easy reach of a wading fisherman, especially during the dark hours. But this was an ill wind that blew no one any good, and by ten o'clock of the second evening we hadn't raised a fish. I had gone back up to the cabin for a late sandwich and was lying on the sofa resting my eyes when Frankie came bursting through the door. It was a singular entrance for Frankie, who isn't the bursting type, and it augured some kind of singular event. "Ol' Coon, grab your fish pole and get down to the dock! We got company!"

From one hour to the next, the wind had lain. A dead calm prevailed and the foam-capped waves that had been crashing on the stony north shore for seventy-two hours had been replaced by a long, slow swell that was growing quieter and easier by the minute as if the lake were sighing before falling asleep. Frankie led the way to the end of the cabin's boat dock, which was lit by a strong floodlight on a twenty-foot pole. He stopped and turned, and pointed down into the black lake. "Visitors, Ol' Coon," he said, with the sly, conspiratorial, we-got-the-bastards-dead-to-rights grin that Frankie wears when he's really got the drop on something—which is frequent.

The lake was five or six feet deep there near the end of the dock with a gravel bottom that was usually visible through the clear water. It was not visible now, of course, because it was night and the water was also cloudy with wind-stirred detritus, but mostly because the bottom was obscured by ranks of fish. They might not have been visible at all had it not been for their eyes reflecting the big dock light. Hundreds of milky opals shown up out of the dark lake; the circle of lighted water was filled with them, the cat's-eyes of night-hunting pike ranked in shoals and banks in the heaving shallows of the lee shore. We could not really see the fish themselves. Only those cold, glowing eyes, some fairly near the surface, others almost on the bottom. They were not entirely motionless; they seemed to shift slightly, adjusting their positions, but always facing inshore as if waiting for something. Frankie and I knew that walleyes tended to school, but neither of us had ever seen anything like this. There was something eerie about those shoals and ranks of cold, jewellike eyes waiting there, especially out at the edge where the circle of light met complete darkness and the eyes seemed even larger and colder. When we flipped our walleye lures out beyond that circle of light and began to retrieve them, we found out what the eyes had been waiting for.

Now, be it known that Frankie and I are not violators of fish and game laws. Or, at worst, very rarely. And we deserve stars in our crowns for staying honest that night. There was no one around, just us and those jillion walleyes. Our subsequent efforts, of course, took us through midnight and from one daily creel limit into another until we each had a carefully high-graded possession limit of select walleyes. We'll always remember that, and the subsequent publicity and the free week at Crandall's Lodge that it earned us. But most of all, those shoals of luminous eyes, watching and waiting like souls in limbo, hanging between heaven and hell in the lake of spirits.

The strange, opaque eyes have a unique structure that provides acute night vision. Many fish can feed in amazingly low light intensity by taste, vibrations, and "skylighting" their prey as it swims between them and the surface. But none are better at this than walleyes, which move into shoal waters in evening and hunt largely by sight during the night. Come daylight—especially if it brings dazzling sun on clear water—the walleyes retreat to darker, deeper water where they may continue feeding along reefs and rock piles during the day, but always returning to hunt the shoal waters in hours of darkness.

Walleyes, like smallmouths, are competent judges of water quality and won't put up with much turbidity and pollution in lake or river. This is probably the main reason walleyes are fish of the Upper River and not of the Mississippi south, broadly speaking, of the Iowa-Missouri line. From there on, the River apparently gets a little too thick for walleyes and they are replaced by a close family member, the sauger, that is far more tolerant of muddy water. The two fish are almost identical except for some minor structural and patterning features, but the big difference to fishermen is one of size. The world's record walleye weighed twenty-five pounds; the record sauger was under nine.

Awhile back I messed up a perfectly good fish story for a Baton Rouge sporting-goods dealer. Upon learning that I lived just up river from Saint Louis and had fished the Mississippi all the way to its source, he proudly produced a frozen fish and said: "Well, then, here's one of your ol' buddies—a real walleyed pike that I hung in the River only forty miles from here!" The dorsal fins had been frozen erect and there was no mistaking the *Stizostedion* persuasion of the twenty-inch fish, but the scattered black spots on the spinous dorsal left no doubt that it was a sauger. The walleye's dorsal is largely unspotted. I pointed this out, trying my best not to be a smart Yankee, and the disappointed dealer took it with good grace—especially when I noted that although walleyes were essentially northern fish, it took a good ol' Tennessee lake to grow the world's record.

River walleyes rarely top off as big as lake walleyes; nor do bass or bluegills or northern pike or just about any other game fish. I can't think it's a matter of fertility and nutrition; could be that the constant stress of current, however slight it might be, has something to do with it. Oh, the average weights of river and lake walleyes may be quite close, but the top figures usually go to lake fish. Still, a good Upper Mississippi walleye is enough fish to distract any real angler from the petty grievances of duty. Hook an eight-pound walleye in the tailwaters of some Upper River channel dam and I'll flat guarantee that for ten or fifteen minutes you won't be brooding about taxes, women, or the federal deficit.

In its primary Mississippi River range, which is roughly from Keokuk, Iowa, northward, the walleye is surely *the* trophy game fish. There are bass and pike specialists who may scoff at that, and I'll admit that I'd much rather hook a five-pound smallmouth than a five-pound walleye, but we are in a lean minority—and probably we should be, for the walleye is not only the biggest game fish in the Upper Mississippi (exceeded only by some northern pike) but is also a superb table fish. No other game fish in the Upper River represents so many pluses to so many people.

Taking the entire Upper Mississippi as a whole, however, the sunfish clan is probably the most heavily fished by angling even though it consists largely of "panfish" that are too small to be technically regarded as game fish—a curious distinction that I have never understood. Excluding the largemouth and smallmouth basses, which are true sunfish that have outgrown the panfish stigma, the Upper River's *Centrarchidae* includes black and white crappies, bluegills, northern rock bass, pumpkinseed, warmouth bass, and several others. Crappies are the largest of these; two-pound fish aren't uncommon, and I've seen a black crappie from the backwaters near Muscatine that went nearly three. A lunker bluegill—a truly big, deep, bison-humped, dark-browed, pugnacious bruiser of a bluegill—may weigh all of twenty ounces. But it makes the most of what it has. It fights. If that bluegill was weighed in pounds instead of ounces, I wouldn't have the courage to ever wet a line in bluegill water.

For years, crappies and bluegills have constituted the top sport-fishery in the Upper Mississippi, although in recent times the numbers of these panfish in total angler harvest have dropped significantly while the take of walleyes, sauger, freshwater drum, and green sunfish has increased. No one is sure why, although the Upper Mississippi River Conservation Committee suspects that more access to the river, with much greater use of motorized boats, has helped increase fishing pressure at wing dam sites and channel dams. Or perhaps panfishing has undergone a decline because of siltation of shallow-water panfish habitats.

Most of these sunfishes tend to be citizens of the backwaters—the running sloughs, island ponds, and lakes away from the main thread of the channel—and may occur there in great numbers. Miller Lake, a backwater near LaCrosse, Wisconsin, is an example. Netting and poisoning in this twelve-acre lake produced over fourteen thousand bluegills, three hundred largemouth bass, and "many" yellow perch, crappies, and bullheads. Counting all types, the total standing crop of the small lake was over forty-five thousand fish—and the biologists on the project believed this was typical of the Upper River's backwaters.

These are some of the majors of the Upper River's scaled fish. There are myriad species of minors as well—darters, carp suckers, quillback, mullet, small minnows, eels, some lampreys, and even the odd trout. But probably more than to any of these, majors or minors, the mighty Mississippi belongs to that tribe of scaleless, misunderstood, and misrepresented river residents known as catfish.

As a family, they aren't the loveliest fish in the crick. Their flattened heads with large mouths and fleshy barbels, the little eyes, the rubbery, scaleless skin, and the family habit of scavenging for food in the silts and slimes of murky rivers all combine to turn off the fancier whose tastes run to angelfish and Atlantic salmon. But to those of us who have grown up with them, catfish are part of home—albeit a homely part, like a beloved aunt who dips snuff and has a wart on her nose.

They have a great size range, and from the first comings of white men to the Great River, giant catfish have stirred the imagination and kindled folk tales.

A fur trader and adventurer named Peter Pond kept a journal of his adventures in the Upper Mississippi valley, and a fragment of that journal—apparently written in about 1765—was found in a Connecticut kitchen in 1868. Peter Pond and his men had some good catfishing somewhere in the Upper River, although he didn't say where:

> We put our Hoock and Lines into the Water and Leat them Ly all nite. In the Morning we Perseaved there was fish at the Hoocks and went to the Wattr Eag and halld on our line. They came Heavey. At Lengh we Hald one ashore that wade a Hundred and four Pounds—a Seacond that was One Hundred Wate—and a third of Seventy five Pounds. The Men was glad to Sea this for they Had not Eat mete for Sum Days nor fish for a long time. We asked our Men How meney Men the largest would give a Meale. Sum of the Largest Eatgers Sade Twelve Men Would Eat ilt at a Meal. We Agread to Give ye fish if they would find twelve men that would undertake it. They Began to Dres it. The fish was what was cald the Cat fish. It had a large flat Head Sixteen inches Between the Eise. They Skind it—cut it up in three large Coppers Such as we have for the Youse of our men. After it was Well Boiled they Sawd it up and all Got Round it. They began and Eat the hole without the least thing with it but Salt and Sum of them Drank of the Licker it was Boild in. The Other two was Sarved out to the Remainder of the People who finished them in Short time. They all Declared they felt the Beater of thare Meale Nor did I Perseave that Eny of them were Sick or Complained.

These were almost certainly flathead catfish, *Pilodictis olivaris*, the largest catfish in the more northerly stretches of the Upper Mississippi and a species that commonly reaches fifty pounds and not uncommonly tops a hundred. Down here around Saint Louis we sometimes hear it called "goujon"—a name increasingly common as one heads downstream into the Cajun country—but more often along the Upper River it is simply "flathead." Its yellowish, olive-drab, sometimes mottled body tends to be a bit slender—at least in proportion to that ungainly, broadly flattened head.

It is a pure predator, not only a meat-eater but one preferring live prey that makes the mistake of happening on the grumpy, solitary flathead as it lurks under a drift pile or cutbank. This is where the hand-fisherman "noodles" for the big fish, wading chest-deep and groping back into the dark recesses until he touches the "goujon," then slowly, carefully easing his hand into the gaping mouth until he can seize a gill arch and haul the fish out into open water. It is a scenario that has never done much for me. I'm not squeamish and have grabbed my share of things that bite and sting, but the prospect of blindly groping back into those tangled haunts of great sullen fish and vicious snapping turtles isn't one that turns me on. There are persistent stories of people having their arms broken while noodling big catfish, or of being dragged back under a deep drift and held there and drowned. They are apocryphal accounts, probably, but thrusting one's arm into the maw of a ninety-pound catfish could produce a certain amount of confusion about who'd caught whom. The last such report I have heard concerned a man gigging fish around the edge of a Kansas reservoir. Late in the day, in shallow water, he apparently found a large flathead catfish. He tied the end of the spear's rope around his waist before spearing the fish—a major basic mistake compounded by another. . . . They found the man's drowned body still roped to the dead catfish. Or so the story goes.

Still, the flathead isn't our biggest catfish. The heavyweight title belongs to the blue cat, *Ictalurus furcatus*, or "fulton" as it is called around my part of the River. In his excellent work *Fishes of Missouri*, biologist William L. Pflieger of the Missouri Department of Conservation notes that in November, 1879, a 150-pound blue catfish was sent to the National Museum by Dr. J.G.W. Steedman, chairman of the Missouri Fish Commission. The good doctor had bought the fish in a Saint Louis fish market, and in a letter to the United States commissioner of fish and fisheries, Prof. Spencer F. Baird, Dr. Steedman noted: "Your letter requesting the shipment to you of a large Mississippi Catfish was received this morning. Upon visiting our market this P.M. I luckily found two—one of 144 lbs., the other 150 lbs. The latter I ship to you by express."

There are stories of even bigger fish. In his book *Steamboating Sixty-five Years on Missouri's Rivers*, Captain William Heckman mentions a 315-pound blue cat taken from the Missouri River near Morrison, Missouri, "just after the Civil War." He commented that during this period "it was common to catch catfish weighing from 125 to 200 pounds" from the Missouri River. Bill Pflieger notes that there are still some sizable blue cats being taken; since the mid-1960s fish weighing 89, 90, and 117 pounds have been caught in Missouri's Osage River. The pole-and-line record is a 97-pound, 57-inch fish.

Blue cats are big-river fish; they don't often range up into the smaller branches and tributaries of large rivers as flatheads sometimes will. They like big streams in general and the River most of all. One of the early common names for this species was "Mississippi catfish," although blue cats are common in many of the large tributaries of the Lower Mississippi and Lower Missouri.

They also tend to be southerly fish, more common downstream than up. Their numbers begin petering out just above Saint Louis and the first of the channel dams, and from there northward they are progressively less common than other catfish. This decline in abundance seems to be linked in some way to the coming of the big channel dams, for the fading of the "fultons" has been greatest in impounded portions of the River. It is not surprising, because the blue catfish is the most migratory of all catfish, going far upstream in spring and back down in fall and winter. The big fultons were once common in the warm months as far upstream as Keokuk, always vanishing in wintertime, and since the advent of the channel dams they have virtually disappeared in the summer as well.

One of the channel catfish, genus *Ictalurus*, the blue cat is a far more attractive creature than the flathead. A pale bluish-silver or light gunmetal in color, the fulton is a well-favored fish, especially in the smaller specimens, trim and streamlined from its deeply forked tail to the modestly flattened head. The aesthetics don't extend to diet, though. Although the blue cat's diet is largely other fishes, mussels, crayfish, and aquatic insect larvae, it may eat almost anything that is dead or alive, animal or vegetable, still or moving, just so long as it is relatively organic and small enough to be ingested.

At the other end of the size scale are the little mad toms, smallest of the catfish tribe, slender fish that are rarely more than five inches long. They are secretive little creatures, hiding by day under rocks and in other cover, emerging at night to range and feed. To the fisherman they are notable for only two things: 1) they can be excellent live bait for walleyes or smallmouth bass, especially in smaller streams, and 2) they have poisonous spines that can give the unwary angler something to remember. Most American catfish have these strong sharp spines—one at the front edge of the dorsal fin, and one at the leading edge of each of the two pectoral fins. In the mad toms these spines have small poison glands at the base, and if one of the spines punctures flesh some of that venom is injected. Many authors state that the result is similar to a bee sting. Well, I've been stung by both bees and mad toms often enough, and I see little similarity. (Although I've never known a prairie boy to have been "stung" or even "spined" by a mad tom or catfish; rather, we are "horned.") Of course, the effect of being horned by a mad tom varies: It is at least a hard bee sting; at worst, it's an almost electric jolt that produces a dull, throbbing ache that may persist for several hours.

I like to think I have a charitable attitude toward many critters regarded as unsavory: skunks, most snakes, lizards, and salamanders, eels, leeches, and even ticks. But there are two I flat don't like and never shall: common water snakes and mad toms. They have often wounded me, and the fact that it was always my fault doesn't change things a bit. Anyway, I take mad toms a bit more seriously than their size might seem to warrant, and I'm not alone in this. Mussel fisherman Worth Emmanuel of Maiden Rock, Wisconsin, noted the invasion of "a hard, rubbery, small-leaved weed resembling ground cover" that had begun to blanket a good mussel bed in the lower end of Pool 9. The plants acted as silt traps, and it wasn't long until parts of the mussel bed were buried under as much as twenty inches of vegetation and silt. Worth worked these beds with crowfoot bars to tear out the plants and put the silt into suspension so it would wash away. Scuba divers might have done that more effectively, but the plant beds held many little mad toms or "willow cats," and Worth and his divers "didn't relish contact with them."

Just up the catfish-size scale from the mad toms are the bullheads—mid-sized, chunky, pot-bellied little catfish that may weigh as much as three pounds but average much less than that. They're the panfish of the catfish tribe, never large, not particularly gamy to catch, and among the unloveliest of fish—but prime doin's in the frying pan. I marvel at how much a fish can resemble Yassir Arafat and still be so eminently desirable. None of the bullheads like heavy current, so in the Mississippi they occur in river lakes, sloughs, and other quiet backwaters. To the rivermen who know them best, they are treasured. I know several old-timers on the River who simply angle for nothing else. Bullheads begin taking bait as soon as the ice goes out and keep it up until the River freezes again, and even in the heavy dog days of late summer, when bullheads from the tepid water aren't supposed to be fit to eat, they are delicious.

And so, dearly beloved, we are ushered at last into the majestic presence of King Cat, the prairie trout, the fantastic fiddler—the lordly channel catfish, *Ictaluras punctatus,* that trim, sagacious, freckle-flanked, clean-lined, fork-tailed, bighearted wonder of a fish—one of the best of all reasons for being a corn-country boy in early summer, or for going down to the River in any summer of one's life.

It resembles a small blue catfish, but unlike the fulton the channel cat usually has spots and flecks along its sides. The back and sides are an olive-brown or brown or slaty-blue tending into the silvery-white belly. The tail is deeply forked, the body is graceful and strong without the burliness of a big carp or buffalo. It is mid-sized, as catfish go. I think I've never seen a channel cat that went over fifteen pounds, and have never hooked one that was more than ten, and that was in a farm pond so I'd never count it as a real, self-respecting channel catfish. The standing pole-and-line record is a fifty-eight-pound giant taken from a North Carolina reservoir. This is about as big as the fish ever gets, and then only in lakes or reservoirs, but comparing such fish to the channel cats of spring-fed, gravel-paved little rivers, or those living in the main thread of the Mississippi's channel current, is like comparing a *sumo* wrestler to a world-class miler.

Channel catfish are simply that: catfish of the channel. They like moving water best and seem perfectly happy in strong current. They are good judges of water quality, as well. I've caught them in Quebec lakes while fishing for walleyes and brook trout, and in spring-fed Ozark rivers where we took equal numbers of small-mouth bass and channel catfish on the same baits.

The channel catfish can make itself at home in turbid water and relatively low oxygen levels, or in high-quality lakes and streams that it may share with trout, although its typical preference is for rich streams that are colored the translucent brown of a Hawken rifle barrel. And although this catfish does well in ponds and lakes, it is first a fish of the moving waters, of the channel current.

Like its big cousin, the blue cat, the channel catfish is an om- nivorous eater that finds food by sight, by touch with its fleshy barbels, by taste, and probably by vibrations in the water. Since it can detect just about any kind of food in any kind of water, the range of its diet is immense. With pole and line I have taken channel catfish with live minnows, frogs, newly hatched sparrows, grass- hoppers, crickets, angleworms, crayfish, fresh chicken guts, sour chicken guts, shrimp, fresh cut bait, sour cut bait, coagulated blood, dough balls, cheese, mussel meats, Ivory soap, and artificial lures including streamer flies, wet flies, spoons, spinners, and deep- running plugs—and I'm sure that doesn't cover them all. In ad- dition to every sort of critter that may occur in a stream, the channel cat may also accept offerings from bankside, including cottonwood fluff, wild grapes, red haw fruits, and other stuff that may drop into the water. There are times and places when channel cats gorge themselves with elm seeds.

Admittedly, some of the channel cat's dietary items would gag a maggot. For years, one of my favorite catfish baits was a com- mercial concoction known as "LineBuster," a doughy mess whose prime ingredient was very bad cheese. It was so bad it was beautiful, and had many applications. A bit of LineBuster smeared on the inside of a friend's leather hatband in hot weather, for example. Or a few dollops of LineBuster spread on the exhaust manifold of a game warden's car. Verily, an unguent for all seasons—and especially for catfishing time.

I love to catch channel catfish on hook and line; they can be wary fish, and are invariably good fighters. During a never-to-be-forgotten evening on Enemy Swims Lake in northeastern South Dakota, when Frankie Heidelbauer and I were fishing with light tackle from a canoe, we took turns fighting a big channel catfish for over two hours until the fish finally tired of the whole thing and broke off our engagement. That fish, by the way, was hooked on a little 1/8th-ounce dressed walleye jig. The two biggest channel cats I've ever caught—nine and ten pounds, respectively—were taken on a 1/32nd-ounce crappie jig and 3-pound-test line.

If anything, though, the joy of catching channel catfish in a good season from good water is transcended by the profound entertainment of eating same. That combination of a highly catholic diet, good water quality, and the discipline of contending with river current produces a firm, rich, sweet flesh that has few equals among freshwater fish.

I love to gig my western friends on this subject.

For the record, I am an ardent trout fisher. I love to lay the good line and address a likely pocket of wild green water with a well-presented fly. I salute the gallantry and uncompromising standards of wild trout, and their tastes in landscapes. I am honored to join them in courteous contest and, now and then, eat them. There is also the perverse joy in coming up from the stream on a sharp mountain morning with ice in the tiptop of my fly rod and a breakfast catch of pan-sized native trout in the creel, walking into camp with its bacon and coffee smell, swiftly frying the trout and savoring them while my cowboy partner says: "Well, eatin' just don't come any better than this!" At which point I enjoy observing: "Delicious! Of course, cutthroat trout can't *quite* come up to channel catfish, but . . ." and if I happen to be with someone like Bob Olive, I duck.

But although the channel catfish has few peers on the plate, it is wholly without honor in some quarters. The stigma imposed on it by Captain Marryat persisted in England, and when it was suggested in the nineteenth century that American catfish be introduced to British waters a wail of anguish emanated from the editorial chambers of *Punch*:

Oh, do not bring the Catfish here!
The Catfish is a name I fear.
  Oh, spare each stream and spring,
The Kennet swift, the Wandle clear,
The lake, the loch, the broad, the mere,
  From that detested thing!

The Catfish is a hideous beast,
A bottom-feeder that doth feast
  Upon unholy bait;
He's no addition to your meal,
He's rather richer than the eel;
  And ranker than the skate.
His face is broad, and flat, and glum;
He's like some monstrous miller's thumb;
Beholding him the grayling flee,
The trout take refuge in the sea,
  The gudgeons go on guard.

He grows into a startling size;
The British matron 'twould surprise
  And raise her burning blush
To see white catfish as large as man,
Through what the bards call "water wan,"
  Come with an ugly rush!

They say the Catfish flimbs the trees,
And robs the roosts, and down the breeze
  Prolongs his catterwaul.
Oh, leave him in his western flood
Where the Mississippi churns the mud;
  Don't bring him here at all!

The same prejudice followed the catfish around in this country. The eminent Dr. William Hornaday, in spite of being prairie-bred, held catfishes in low esteem, and once said of the blue cat that "even when alive and in good health, it is a very ugly fish—heavy-paunched and mud-colored. It looks like a fish modelled out of river mud." With evident relish, Dr. Hornaday cited an item from the San Francisco *Evening Bulletin* concerning the introduction of catfish into California, where it was not native: "Then the fish commissioners made another unfortunate experiment . . . they introduced the hated and almost worthless Catfish to the waters of California. It was reported, in answer to protests made at the time, that only a superior kind of Catfish would be introduced, against which there could be no objection. But they turned out to be the same old toughs that have occupied western rivers and bayous to the exclusion of better fish. Their value is so low that very few seek them. The Chinese sell them occasionally, as they do carp, if they can find a customer. But most consumers turn away from these fish in disgust."

Things have changed for the better. The catfish is no longer a dirty word in California, and on a national level customers are paying around a quarter-billion dollars annually for dressed catfish. Catfish farming is big business in some parts of the South, dressed catfish are being imported from Brazil, and channel catfish is still the main money fish for commercial fishermen on the Upper Mis-

sissippi. While catfish rank third behind carp and buffalo in terms of weight harvested, they rank first in money value. Fresh from the River and still "in the round" an undressed channel catfish may bring 65 cents per pound, or more. Catfish fillets may be $1.50 per pound—if you can find them.

Oddly enough, there isn't a great deal of pole-and-line fishing for channel cats in the Upper Mississippi. At least, it isn't comparable to angling for walleyes, bass, and crappies. In my experience, the bulk of channel catfishing is done in the "inland" rivers. My old friend "Three-Finger" George Kaufman, veteran game warden at Lansing, Iowa, once told me, "In my territory the main River ranks last as catfish angling water. Very few sport fishermen are working channel cats out there, probably because there's just so damned much water that it's hard to locate the fish. Besides, there are lots of other fish that are easier to find and easier to catch."

Part of the problem may be that channel catfish are inclined to head into deeper water during the day, coming up into relative shallows in the Mississippi only at night. And while this can complicate matters for the daytime fisherman, it simplifies things for an island-camper with a taste for fresh channel catfish and some likely shoal water for his all-night bank lines. Add a black Labrador for companionship and a touch of comic relief, with a driftwood breakfast fire in the cool sweet morning and the first sun touching the high stone battlements over there across the channel, and we've come full circle.

It's time to kick out the fire and call the dog. Daylight's burning. Time to join the River.

4

RIVER YEAR

Spring

Even for early April it was a wild morning.

I had pulled out of Clinton, Iowa, with a headful of plans, the pickup truck full of walleye-fishing gear, a Willie Nelson tape in the deck, and the new johnboat trailering smoothly behind. It had been windy in town with overhead traffic signals swinging crazily and streetlight standards swaying; I noted all this in passing, but with the Bellevue Dam and huge spawning walleyes in mind, what matter any swinging traffic lights?

U.S. 67 climbed north out of town, and up on the flat, exposed highlands I began to feel the press of the west wind that had

followed the rain of the night before. I remember glancing down
at Pool 13, one of the most treacherous impoundments on the
Upper Mississippi—a river-lake almost four miles wide and over
six miles long, open to the full sweep of the wind—and seeing the
whitecaps raging across it, and thinking how mean it was down
there and what a bad place to be, when there was a grinding crash
behind me. A seventy-mile-per-hour gust had caught the unloaded
johnboat, heavy as it was, and flipped boat, trailer, and motor
upside down in the middle of U.S. Highway 67.

The trailer hitch had never torn loose; it was still locked faith-
fully on the ball, while the heavy trailer tongue was twisted 180
degrees. Otherwise, the trailer was almost unscathed, but the boat's
starboard gunwale was mashed flat and part of the bow had been
smeared along the pavement. Back down the highway were strewn
several principal parts of the big outboard motor, which had been
torn off the transom. I leaned against the raw April wind roaring
unchecked across the open fields to the west and absorbed the
scene with considerable pain. I'm hardly a flaming materialist, but
the outfit was less than three months old and had stood me a hard-
earned $3,400, and for a while there I was about half sick. Maybe
even a little more than half. On tape, Willie Nelson was singing
about blue eyes crying in the rain.

By this time several pickup trucks had stopped and a motley

squad of farmers and rivermen helped put the trailer back on its wheels and off the pavement. A deputy sheriff arrived, made out a report, and radioed to Camanche for a wrecker that could haul the mess away. On his advice it was trucked to a Clinton marina, Lateke Sportscenter, and turned over to service manager Larry Kahler. As it turned out, that was about the only lucky break I got that day. Larry did a masterful job of straightening, welding, and motor repair, and although the outfit emerged from the experience with a much saltier look than before, it would prove to run better than ever. The insurance would cover almost all of it, but there was a lonely interim when I had neither boat nor check.

In the meantime I was a shorebound walleye fisherman whose favorite spots were well offshore. I would be on the beach without the big johnboat for almost a month, meaning there would be no April walleye fishing below the channel dams, and I would miss a vernal rite of great significance on the Upper Mississippi River.

Through the summer walleyes may be widely distributed through much of the River between channel dams, although more than likely they're in deepish water around rocks or bridge structures or off wing dams most of the day. With the cooling days of fall, however, they begin to move upstream in an annual pre-spawning run. Most never leave their home pool—their particular stretch of Mississippi between channel dams—moving upstream into the deeper, well-scoured channel that extends for several miles below a dam, and almost all the walleyes in that particular "cell" of Upper Mississippi may hold there for the winter. They can be caught then even though their metabolisms are at a low ebb and their feeding is desultory. But they are eminently worth the catching. All summer they have been lazing over deep gravels and glutting themselves on forage fish and crawdads, putting on fat and muscle, and at their arrival on the winter holding beds they are prime and solid. The trouble is, fishing for them from an open boat in the Upper Mississippi's channel in late November can be as dangerous as it is bitterly uncomfortable, and the action is usually as slow as the fishes' feeding. Then, too, you're pestered by the nagging concern that you're missing out on some good hunting. I have never fished walleyes in the Upper River in late fall without regretting I was there at the time, and being glad later on that I'd gone. Even with a bucket of glowing charcoal amidships, and wearing as many layers of wool and down as I could get under the life preserver, it's a bitter business. But afterward, with one or two good walleyes that were heavy with that white, flaky meat, I always rejoiced in having gone.

The real action, though, is the spring rite that comes soon after ice-out, with water temperatures rising to about 45 degrees or more. In the prime walleye region of the Upper River from about Bellevue, Iowa, up to Hastings, Minnesota, this all begins in late March or early April and lasts about three weeks. Just about the length of time my boat was in dry dock.

The little male fish move upstream first, often as soon as the ice has gone, and are soon followed by the big females now heavy with spawn. Most gather at the edge of the main current, holding over rock and gravel near the ten-foot contour in the vicinity of the big dams. At this time, as much as 90 percent of the walleyes in a thirty-mile stretch of Mississippi may be concentrated in the half mile just below the upstream dam. Years ago, using electro-fishing gear in an April study below the Guttenberg, Iowa, dam, our survey crew took 141 walleyes the first night. In the five days that followed (five nights, rather, since many of the fish were too deep during the day to take with shocking equipment) we took over 1,150 walleyes and saugers that were tagged and released. Most of these fish were turned in rocky or gravelly shallows in late evening, the majority within a few hundred yards of the dam. I can't recall that more than eight or ten fish were taken over a thousand yards below the dam.

With spring walleyes come spring fishermen. The fish are concentrated now, and although the weather can still have a wire edge, it doesn't cut to the bone like that December wind. Some people fish from the banks as close to the dam gates or lock bays as they can get without being run off by the lockmaster; others are in boats, often as close to the dam as the law allows, working baits of big minnows in conjunction with spinners or weighted jigs just over the rocks and gravels of the riverbed. The action can be furious, and for a couple of weeks the fishermen may be in hog heaven. One of the master anglers of the Upper River, outdoor writer Dan

Gapen, luxuriates in the memory of the 1977–1979 spring fishing below the Red Wing dam when lunker walleyes were hitting so hot and heavy that the Department of Natural Resources closed limited areas to fishing. "For years this area has been a hotbed of spring walleye frenzy," Dan reports, "though no one understands exactly why. It's not the only good spot, either, because large concentrations of walleyes are known to spawn along wing dams downriver from Hastings and at the Hastings dam itself."

It's usually all over in two or three weeks. Spawning done, the walleyes begin breaking away from the dam site and dispersing downstream. A week of so after that I collected my boat, trailer, and motor—but nothing could repair my totaled walleye opening.

Spring is gaining strength; in the uplands above the River in northeastern Iowa, Wisconsin, and Minnesota, and even in some bottomland timber, ruffed grouse cocks are on the drumming logs and beginning their spring tympany, beating their cupped wings against their chests in the staccato *Pud-pud-pud-pud-pud-pudpudpud!* that always reminds me of the old one-cylinder gasoline engines once used to power washing machines on back porches. Geese and mallards are moving through, with canvasbacks and tundra swans cutting across the Upper River on their long journey to the prairie provinces and Alaska, the tender teal beginning to come as well, and wood ducks finding the snags and hollow staubs in the backwater lakes.

Bald eagles are moving upstream as the ice goes out, and on backwater lakes up around Trempealeau, Wisconsin, commercial fisherman Ted Koba and his crew are seining, discarding carp suckers and gizzard shad with the huge eagles coming to within a few yards for the bounty, like park pigeons eating breadcrumbs. On an early April morning soon after ice-out you can hardly travel along the Upper River without seeing eagles. Some are barely visible motes, incredibly high; others are perched on blufftop snags or in lofty riverbank cottonwoods. And maybe, if you're favored by the red gods, you will see a courting pair of bald eagles dancing high above the River, their talons clasped as they whirl and fall in unison. Other raptors are up there, too, the red-tails and red-shoulders and harriers, northering in wheeling circles that seem pointless and unproductive until you suddenly realize they have moved across the whole vault of sky as you've been watching—going around in circles but getting somewhere.

Some sheltered, reedy bays of the vast Wabasha Bottoms north of Winona may still be frozen in late March and early April, and sheets of waterfall ice persist on the rocky walls of highway cuts. Within a few weeks most of the geese have passed through, but the big backwater lakes and main pools of the Upper Mississippi are clamorous with courting, restless wildfowl. The River is high and some low cropfields are inundated. I worked down a steep, grassy bank and lay in deep grass out of the wind one early April day, watching the show on a broad, shallow bay that had drowned an unpicked cornfield. There were rafts of lesser scaup placidly breasting the choppy water, letting the brisk northwest breeze blow them across the open flats until they neared the edge of the flooded corn, when they would turn into the wind and make their running takeoff, flying to the far side to begin another downwind drift—apparently half asleep. Where the flooded cornstalks began, the mallards began. I could hear their steady chuckling whenever the wind lay for a moment, and although not much could be seen because of the low angle there was certain knowledge that the unharvested field was, as Glen Yates would say, "black-full of poultry." Scattered through the edges of the corn and out in the open shallows was a host of shovelers, ring-necked ducks, green-winged teal, red-heads, wigeon, and a few canvasbacks.

Later that day I stopped near Fort Madison at a place overlooking the vast, open sweeps of Pool 19. It was a spring waterfowl mecca. I clamped the 48x spotting 'scope on the door of the truck and glassed the portion of River within my range of vision—a gray, windswept segment two miles wide and several miles from side to side—and from any position there were always ducks in the field of the telescope. They were almost all divers, blithely riding the wild chop impervious to cold and wind. Many were asleep, their heads underwing. Most were lesser scaup in great rafts spread on the open pool like tattered fabric, here tightly knit, there open and torn. With them, but apart from them in their own bunches, were redheads and the lordly canvasbacks—the *canard chevals* or "horse ducks," named for their straight, thoroughbred profiles.

Some of the birds were resting; a few were engaged in nuptial
flights. Many of the "cans" and redheads were feeding, diving
deeply and reappearing briefly and then diving again. Some were
nearby, but even with the powerful 'scope I could not detect any
rootstocks or other plant materials in their bills when they surfaced
after a dive. They were obviously battening on Pool 19's blue plate
special—the fingernail clam. This fingernail-sized mussel is a succulent bit of rich nutrition relished by the diving ducks. It occurs
in vast numbers in Pool 19; there are places with up to ninety-four
thousand of the tiny shellfish per square meter. Diving ducks may
eat twenty-five hundred tons of these mussels in this one pool
during a single fall migration. On the basis of a "duck-day" (one
diving duck using the pool for one day) Frank Bellrose of the Illinois
Natural History Survey figures that this Keokuk Pool has about
twenty million diving duck-days annually—the most important inland water for diving ducks in North America.

The beds of fingernail clams are only part of the pool's rich
macroinvertebrate population. Oligochaete worms may number up
to fifteen thousand per square meter, with myriad snails, mayflies,
leeches, and my old research subjects the chironomid larvae—the
semitransparent young of phantom midges. For years this great
river-lake has ranked within the top six of the Upper River's twenty-
five pools in commercial fish harvest. Channel catfish, prime money
fish of the River, dotes on fingernail clams just as ducks do, and
Pool 19 has produced the biggest single modern channel catfish
harvest of any pool—657,616 pounds in a single year.

Somewhere over there near the Illinois shore, set in the shallows, I knew there would be several duck traps baited with corn.
The boys would be baiting them and running them now—Illinois
wildlife biologists George Zebrun and Dave Harper and their aides—
checking the open traps several times a day for ducks that had
been lured into them. I grinned, thinking about it. They had rolled
out of their warm bunks in the hillside cabin early that morning
and gone to the River. There would be canvasbacks and redheads
and scaup milling in the traps, and my friends would slip into their
patched waders and slowly work out to the traps, sinking into the
flocculent silt that is known commonly as "loon shit" to biologists,
and shipping water over the tops of their waders. Dave Harper

yells: "I think my butt is frozen but it's too numb to tell!" And George Zebrun, the Mad Russian who recently transferred to Wildlife after years in the Fisheries Section, replies: "Aw-w-w, is him wet and cold? Candy-assed wildlifers!" Zebrun is no stranger to icy water. But neither is Dave, a longtime trapper of beaver, and The Russian knows it.

The ducks would be crated and taken back to the cabin to be banded, their species, age, sex, weight, and identification numbers entered into the official record before their release. Tossed into the air toward the River, they depart with flickering diving-duck wing-beats, invariably expressing their opinion of the whole bureaucratic process by defecating on the way.

A duck band is a ticket in a lottery. Next fall a hunter may find the band on the leg of a duck he has just shot and if he reports that band number and the location and date of the kill, he will be rewarded with information about where and when the duck was banded, and how old it was at the time. One of the indignant canvasbacks released there at the cabin on Pool 19 may be shot next November on Chesapeake Bay, or three years later near a staging marsh in eastern Saskatchewan. It is best to know as much as possible about the times and travels of that duck—and a thin aluminum band can tell it.

April, and the wood ducks are in and scouting for nest sites.

We came fearfully close to losing this gorgeous little duck, largely because of its nesting habits. Unlike most other ducks that flew beyond the heaviest spring hunting to nest in the thinly populated northern prairies, woodies nested almost entirely in the eastern half of the United States, where they were subject to shooting pressure during their entire breeding season. There was direct hunting mortality as well as harassment during nesting time, with subsequent disruption of pair bonding and normal breeding patterns. As if that weren't enough, the wood ducks depended on the hollow trees and snags along streams and rivers—and widespread timbering and bottomland clearing were destroying countless nesting sites. At about the time of the First World War the wood duck was rare and growing rarer, and some feared that woodies would never survive the 1920s.

They were saved by prohibition—not the Volstead Act kind,

but by a Migratory Bird Treaty Act between Canada and the United States that put an end to market gunning and spring shooting. Within a decade the "summer duck" was out of danger and doing well. Protection had much to do with this comeback, but that alone is never quite enough. It must be leavened with some positive, hands-on management, and the wood duck was a natural for this—presenting one of those rare and classic situations in wildlife conservation when everything comes into focus. This was something that the outdoor public could readily understand, sympathize with, and really do something about. Wood ducks needed housing, and would readily accept man-made nest boxes mounted on streambank trees or on poles above water. State and federal wildlife agencies began making thousands of such nest boxes and were soon joined by sportsmen's clubs, landowners, Boy Scouts, Audubon Society chapters, and just about anyone who could handle a hammer and saw and had the welfare of the wood duck at heart. Ah, for the sweet, lost days when private citizens could become so personally involved in practical game management, with such gratifying results. Today the wood duck is one of the commonest waterfowl along the River, and going stronger than ever.

I see them as early as mid-March, house-hunting in pairs among the river birches and red-budded silver maples, or started from some hidden pond in the bottomland timber and flying headlong through the trees, twisting and dodging like ruffed grouse, the duck crying an alarmed *wee-e-e-e-k, wee-e-e-e-k!* to her silent mate.

Except for woodies, the marshes and backwaters of the Upper River are not notable waterfowl producers these days—not what they used to be, anyway. Some mallards and blue-winged teal still nest around the marsh edges, but most wildfowl head on over into the prairie regions of potholes and marsh lakes. What the River lacks in duck production, though, it makes up for in herons. Night herons, both black-crowned and yellow-crowned, bitterns or "slough pumps," the great and snowy egrets, little blue heron, green-backed heron, great white heron—but most of all, on this heronish river, the great blue, old Uncle Spindleshanks himself, a fish spear with an attached bird as a life-support system.

This is the biggest bird on the River, with wings spanning seven feet, and a long neck and stiltlike legs that put the seven-inch stiletto of a beak four feet above the quiet shallows where the great blue heron stands motionless waiting for fish, frog, or whatever to blunder within range. Beware that lightning stroke that can drive the beak through the blade of a spruce canoe paddle. A captured heron may try for a person's eye; I know a conservation officer whose glasses deflected such a blow, giving him an ugly gash in his forehead. But a dagger stroke from a captive great blue heron isn't the only peril of close acquaintence. I was once drifting down a looping backwater near Cap au Gris just south of the Winfield Dam, power off and letting the light breeze and current move me while I leaned back against the outboard motor almost asleep. The boat was rounding a blind turn in the running slough and was more or less hidden by the rootwad of a storm-blown cottonwood and parallel with the tree trunk before the heron saw me. The big bird had been standing on the log when I appeared almost at his feet, and there was no way out of there except right over my bow. Startled herons usually void excess baggage for a quicker getaway, and this one put a rope of white excrement across the bow of the freight canoe. It was a mess, but a good charge of comic relief, at that. Nothing is a more uncoordinated set of angles and gangles than a great blue heron trying to get it all together in a panic takeoff; it's something like a wooden camp chair being folded in a great hurry. There is just no graceful way it can be managed.

The big herons may appear on the Upper River with the first hint of spring, well before the main wave of waterfowl. We've seen them just north of Saint Louis in the earliest part of March. They must begin nest-site selection almost at once—a choice that doesn't reflect any maverick individual tastes, for the birds nest in dense colonies with special requirements. Along the Upper Mississippi the main requirement appears to be very tall trees growing in riverside jungles. And although I have seen great blue heron colonies nesting in groves of thirty-foot box elders in the Flint Hills of prairie Kansas, I don't think I've ever seen the big herons nesting along the Upper River in trees that were less than eighty feet high.

These are invariably towering sycamores and cottonwoods, huge trees, some with trunks five feet thick or even more, soaring above the rest of the floodplain forest canopy to well over a hundred feet. The rickety nests of the herons are in the very crowns of such giants—untidy, slapdash platforms of sticks that may occupy twenty acres of the forest's canopy. Far below the lofty, sunlit upper levels is the fetid understory of bedstraw, poison ivy, catbrier, and rotting tangles of driftwood that discourages human intrusion. There must be great blue heron colonies that are easily accessible and relatively pleasant for people to visit, but I've never been there. In any case, a "ripe" heron colony with newly hatched young is usually a mess. The ground below the nests is whitewashed with excrement, bits of rotting fish, smashed eggs that have fallen out of the slapdash nests, and even dead nestlings that have done the same. The birds are sensitive to intrusion; you'll never sneak undetected into a heron colony. Still, in a ripe rookery they abandon much of their usual caution in their concern for the young birds and do not leave the colony, but wheel above the treetops yelling in alarm and irritation—and without a puckered sphincter in the bunch. It can be pure bedlam at this point, a cacophony of whoops, harsh clucks, and a particular noise like rusty gates. So you stand quietly then, hoping to not be caught in a drizzle of defecation and partly digested fish and frogs, and trying to ignore the gray veil of mosquitoes that sings around your head. Soon the great birds begin calming, floating airily back down to their nest-perches, and the colony resumes the muted, steady clattering like a flock of blackbirds in a distant marsh.

The colony is never empty of adult birds; one parent or the other leaves and returns, sometimes hours later, with lower neck distended with the animal stuff that is puked into the maws of the greedy nestlings. The approach and landing seldom vary. The old heron circles the rookery and floats in toward the nest, back-stroking to kill-flying speed and settling onto the landing limb as weightless as gray smoke, then furling the huge wings in several neat gestures before sidestepping deliberately along the limb to the nest.

Late April and early May is a maddening time of year on the Upper River. There is too much to do being compressed into too little time to do it all. There are the heron colonies to visit, bull-heads and early bass to catch, bedding bluegills starting to take fly-rod poppers at the surface, and, as the weather warms and the new oak leaves become the size of squirrels' ears, the morels begin coming on.

Out of the fine vernal confusion of an Upper River spring, there emerge certain gustatory perqs that may not be unique to the region, but which are notable among the world's richest blessings. Counted prominently among these are such treasures as wild strawberries, fresh bluegill fillets, walleye and catfish chops (the firm, scalloplike "cheeks" taken from males in breeding condition)—and the morel mushroom.

Maybe I'm more stomach than soul, but to me one of the best reasons for spring coming at all is this wrinkled tan cone that grows in the damp molds of a bottomland timber. There's nothing that can brighten eyes and lighten your step like a mess of butter-fried morels and fresh fiddler catfish. It's Ma Nature's reward for enduring an Upper River winter.

The most common morels are about four inches high and look like wrinkled, deeply pitted crosses between corncobs and ice-cream cones. That general similarity to a new corncob can be striking; many times, out mushroom hunting, my eyes have snapped into sudden focus on a corncob left under a squirrel's den tree. No other mushroom is so easily learned, and that's one reason for the morel's popularity. Supremely delicious, it can't be mistaken for any poisonous species. Most people don't trust themselves to know such fine varieties of fungi as oystershells, puffballs, sulphurs, and inky caps. But once they know the morel, they never forget it. Deeply pitted and honeycombed, and tan or light brown, the cap is textured somewhat like tripe. The cap is usually about three inches high, on an inch or two of whitish stem. There are false morels with broad, wrinkled, brainlike caps, some of which are edible. But they have only a casual resemblance to the true morels, and in years of mushroom hunting I have never found the two types growing near each other.

A big part of morel hunting is calculated luck, but knowing when and where to do it cuts the odds to an acceptable minimum. The best time is May, and the best May is warm and wet. Some of the river folk claim morels can grow to full size overnight in the right sort of weather, but that's doubtful. It probably takes several days for a morel to reach its full, stately four or five inches. No one is really sure; few if any morels have been grown in captivity.

I've always preferred deep, shaded, timbered valleys drained by small creeks for serious morel hunting, or, to break it down farther, in timbered places where May apples grow. More often than not, if there are May apples in a patch of woods there will be some morels not far away. They seem to like the same sort of country. Some morel hunters swear by stands of cottonwoods, or dead elms, around big old willows near riverbanks. And every now and then, when things work out just right, a big river island will produce a morel bonanza.

The wildest mushroom stampede I've ever known was during an unseasonably warm May with much rain. Earlier high water had inundated some of the big river islands and then receded, leaving beds of heavily shaded silt under the greening willows, cottonwoods, soft maples, and river birches. Three boys stopped off on one of these islands during a fishing trip and found the rich silt thick with young morels. It was the first big mushroom strike of the year in an area that reveres morels, and the boys had sold nearly eighty pounds at a dollar a pound before their secret leaked out. The rush was on.

Boatloads of mushroom hunters converged on the islands. People got lost in the dense willow "slaps" but didn't really care. Some tried to mark their trails with white rags, and when that failed they climbed trees and hollered for help. Everyone was joking, yelling, lost, happy—and picking morel mushrooms with both hands. No one will ever know how many loads of morels left those islands, but some hunters reportedly covered the beds of pickup trucks with the little sponge mushrooms.

That sort of thing can happen in many areas, come May, but for some reason the morel's popularity is often highly localized. In places where these mushrooms are not part of the local spring tradition, people may ignore them even when they're abundant. None of this applies to any place along the Upper Mississippi that I know of. Hunting morels is a spring rite along the River, and some small communities will even close shop when the word goes out after a warm spring rain.

By all odds, morels are at their best when they are fresh from the woods. I like to eat them the same day they're found, splitting them in half and soaking them in cold saltwater to wash away the small ants, gnats, and beetles. The mushrooms can be dipped in whipped egg, rolled in pancake flour that's been seasoned with celery salt, and fried in butter. Or fried with steak, or minced into gravy or barbeque sauce. Several years ago I hit a mushroom jackpot and gathered more than five pounds or morels in an hour. That night my wife Dycie and I had sliced morels and T-bone steak, which was very fine. The next night we wondered about trying them with pork chops or scrambled eggs. We debated the point feebly for a couple of minutes and said the devil with it, melted a pound of butter in the pan, fried nearly three pounds of morel mushrooms, and ate them at one sitting. You can always eat pork chops and eggs. May and morels come only once a year.

The far travelers come twice, following the great migration corridor of the Mississippi from end to end—and beyond.

Some of the smallest come farthest, like warblers up from the highland jungles of Mexico, Guatemala, Nicaragua, Columbia, and Peru, and there are finches, vireos, and thrushes from deep in South America. For small birds that must carefully budget their strength and time in the vast journey from one hemisphere to another, the Mississippi valley is a kindly route. It teems with plant and animal bird foods of all kinds; there are no geographical extremes, no high divides or mountains or deserts; it flows easily in a direction that couldn't be better for north-south migrants, and is a compass arrow for those going up into the boreal and subartic regions. The river woods are alive with a multitude of warblers, some staying to nest and others just passing through, and all intensely busy as only warblers can be. On the uplands the lord-god birds, the crow-sized pileated woodpeckers, are cutting rectangular feeding holes in dead snags—the only woodpecker to cut a square hole. The eagles and main mass of waterfowl and shore birds are long gone northward, but the river marshes are alive with male redwings flashing their red-and-yellow epaulets—a sure sign that largemouth bass and bluegills will be rising to topwater lures.

It is a bittersweet time. I drove through Lansing, Iowa, after a visit with my old friend John Spinner, seeing the sheer face of Achafalaya Bluff south of town and knowing that the falcon nest we once roped down to is gone now. Somewhere peregrine falcons have outlasted DDT, but not here. It soured the May morning. The River may be dropping, and the big boat and motor in good order again, with morels coming in and the snail on his thorn, and so forth—but right then I'd have traded it all for the sight of a peregrine perched on a cliff-edge cedar above his aerie. One of my sour old-timer-remembering-when moods was coming on, and it wasn't helped any by the sleek pavement that has replaced the rock road that once ran from Lansing to Harper's Ferry. I stopped at the old Wexford church to look at the creek; there was a parking lot there now, and a sign proclaiming Wexford Creek to be a public access trout stream. I like it a lot better when there were no signs and more trees, remembering an early spring morning when the old priest was at the creekside below his church with the hem of his cassock soaked with dew, flyrod in hand, praising God and blessing His trout.

At Yellow River Forest headquarters I learned that Jack McSweeney had been called to Des Moines for a staff meeting; even here, in the Yellow River country as far from the Main Office as he could get, the planners and bean-counters and paper-shufflers still managed to find him. In a fine growly temper by this time, I drove down to Little Paint Creek and parked, putting a water bottle, crackers, and sardines into the day pack and striking off upstream. As always, instant tonic. Before I was out of sight of the truck, my February mood had begun to thaw.

Little Paint Creek is a stream of many parts. It is rich water, supporting a rich fauna. Although not so gin-clear as an idyllic trout run might be, it is nevertheless highly productive limestone water that drains a rich watershed, and it holds rainbow and brown trout worth close attention. At the moment the sun was too high and bright for serious fishing, but it was as good a time as any to prospect farther upstream than most fishermen go—and maybe run into some morels as well.

An hour later I was several fishing holes wiser and a dozen morels richer when a steep outcrop forced me up the hillside and away from the stream's edge. Finding several more mushrooms and knowing that the spores often wash downslope, I worked up a shallow drainage in the long, heavily wooded hillside and finally topped out beside a dense stand of sumac. I had paused there to catch my breath and get my bearings when something gleamed back under the sumac. It was for only an instant, but when you're alone in the woods and tuned in as you should be, an instant of something unusual or out of place is enough. There was nothing in those woods or in that sumac thicket that would gleam—not at midday with the last of the dew gone. Yet—something there. The smooth shine of a blacksnake? Maybe. Then that electric thrill, the visceral jolt that comes when unexpected marvels are suddenly revealed. The day stopped; the earth stopped. Nothing breathed or moved, all life hung between one pulsebeat and the next. Even the air had changed; there was a hush to it that had not been there an instant before, as if it were a green-and-gold veil holding back silent, waiting sounds. The fleeting gleam again. It came from a half-closed eye less than ten feet from where I stood. With it a form began to materialize, a rich bay broken with patterns of white spots, blending perfectly with the sun-dappled floor of the nursery thicket.

One of my long-standing laments springs from the fact that goblins and elves have lost out in most of our modern forests, exorcised by the superior magic of the cathode ray tube and microprocessor. The Science that has banished them has failed to replace them with anything else as good. Still, there remain among us certain wood sprites that defy exorcism. Not only have they survived Scientific Inquiry, which is something goblins and elves could never do, but they have even been sustained by Science— partly because they have succeeded in enchanting the Scientists themselves. One of these wood sprites lay half-curled on the thicket litter almost at my feet, a white-tailed deer fawn no more than two days old.

It was very small, far smaller than most people think, six or eight pounds of large eyes, ears, and slender legs folded close to the spotted, honey-colored body. It was entirely motionless, not seeming to even breathe, eyes half-closed and ears back against its neck. I was trying not to stare, for staring is very rude in such circumstances. Wild things do not like the direct stare of men, but since I was surely the first man the fawn had ever seen, odds were that I was disturbing its mother far more. She wouldn't be far away. She'd have spent the night with her fawn, of course, and at this time of year, with rich browse everywhere (including morels for which she and I were competing) there was no need to go far for forage. She'd have settled the fawn in the sumac thicket not long before, I guessed, doing whatever it is that mother whitetails do to hold their fawns silent and motionless until their return. She was surely watching me now, had indeed heard me and winded me as I came along the creek and up the long hillside, straight to her wood sprite. It must have been bad for her; she must have been sure I knew exactly where the fawn lay. If a dog or coyote or bobcat or almost any other creature had done that she might have made her presence known, might even have attempted a defense or diversion of some kind. But not with a human—not with most-dreaded, bloody-handed man. So she kept her secret distance while I paused before her fawn, watching, I am sure, in dumb terror.

Over the years I have come across a number of whitetail and mule deer fawns lying quietly in nursery situations of some kind, and have yet to see a doe in attendance. Such are the motherless fawns that people love to rescue. Even if they don't "save" the fawn they can't resist touching it—and drenching fawn and nursery with their scent in the process. I've no thoughts at all about whether or not a doe might abandon or neglect such a fawn, but I do know that coyotes and wild dogs are fond of trailing humans to see what will turn up, and it would not do for such a trail to lead to a young fawn. I had already come too close and stayed too long. I got out of there, backtracking for a hundred yards before continuing along the hillside.

The day had taken a special turn. Anything was possible now that spring had been given the special endorsement of wood sprites. The magic lasted on into the evening, with a pair of ten-inch rainbow trout and a mess of butter-fried morel mushrooms for supper. A couple of whippoorwills were having a hollering contest not far up the valley as I rolled into my sleeping bag, and I fell asleep with fawns on my mind.

Summer

From the end of the long jetty at Portage des Sioux, where the figure of Our Lady of the Rivers towers forty feet above Pool 26, Catholic prelates have duly blessed the fleet. The devout and the curious were gathered in houseboats, cabin cruisers, runabouts, sleek motor yachts, sailboards, and those doughty working craft of the Mississippi, the ubiquitous johnboats. It was the officially sanctified, if not actual, opening of the boating season on the Upper River.

There will be a constant recreational traffic from now until freeze-up. Some of it will be passing through from one end of the River to the other—sinewy canoeists spending months on the adventure of their lifetimes, houseboaters or motor yachters running down to Saint Louis from Minneapolis or vice-versa. I've seen fifty-foot sloops with masts unstepped, running downriver on their auxiliaries, bound for blue water and probably grooving parts of the channel bed with their keels. One morning I passed a small tugboat, an actual hemp-bumpered Little Toot chugging upriver festooned with lines and the morning's laundry, and a weatherbeaten Tugboat Annie type waving happily from the fantail.

I never see visiting motor yachts with "Port of Saint Paul" or whatever gilded on their big transoms without thinking of the cloudy morning I met the Rich Professor.

Doc Kozicky and I had been out since first light that Sunday morning, pulling basket traps and moving them to new sets. A number of these catfish traps had been mudded in, and the process of disinterment was as messy as it was laborious. By the time we pulled into Pére Marquette Harbor we were slathered, smeared, clotted, and crudded with river mud of the foulest and most mucilaginous kind.

We tied up to the harbor wall not far from an elaborate cruiser whose transom proclaimed that home was Ames, Iowa, which happens to be about five hundred river miles from where we were at the moment. It also happens to be my home town. As I walked by on my way to the truck, a well-barbered gentleman clad in shorts and boating moccasins emerged from belowdecks with coffee cup in hand.

"Good morning," I offered.

"Morning."

"Hail from Ames, Iowa, eh?"

He sipped his coffee, appraising my patched, mud-smeared oilskins over the rim of the cup.

"Um."

"Come down the Skunk River, did you?" A little levity, there. It didn't take.

"Hardly. We berth at Davenport." He turned his gaze to the hills beyond the River. I was beginning to think that our acquaintance would never blossom. I was also beginning to enjoy myself.

"You in business in Ames?"

"No. I'm a member of the faculty at the University there."

"Well. What college?"

"Science. Professor of biology." He turned to go below. This was better than I'd hoped for.

"How about that! See my partner over there? Dr. E. L. Kozicky, former professor of biology at Iowa State. Helluva small world, ain't it?"

He bent a horrified glance on the subject of my comment. Doc was standing thirty yards away, oblivious to the conversation and looking even raunchier than usual in his filthy oilskins, lighting his sulphurous old pipe. The Rich Professor gave an audible gasp and dived below. My immediate hopes for a cup of hot coffee vanished with him. I walked over and joined the Poor Professor.

"What was that all about?" Doc asked. I promised to fill him in if he'd buy coffee at the Blue Goose Tavern, and he enjoyed my somewhat embellished account so much that he popped for a cinnamon roll, too.

Most of the boating traffic on any part of the River, though, is local stuff. And most of that is pure joyriding back and forth between the dams in fast little runabouts, stately houseboats, powerful motor yachts, and neat cabin cruisers. There are boats-with-a-purpose such as water skiing or bass fishing, and always, of course, the big aluminum johnboats with olive-drab paint that are out at dawn running nets, traps, or trotlines and are back at the fish market by the time all the joyriding begins. On summer evenings and weekends houseboats anchor behind islands out of the main current for a leisurely round of eating, drinking, loafing, fishing, TV ball games, or just baking in the hot sun. Shore parties establish beachheads on the big sandbars and on certain islands, basking and swimming and barbecuing. On pools broad enough to give the wind a good fetch, there'll be weekend sailing regattas. Now and then, Authority makes a brief appearance in the form of Coast Guard patrol boats and we are all happier knowing that our Cornfield Navy is nigh. And always, through this confused and confusing fun fleet, the huge towboats with their strings of barges churn doggedly up and down the main channel.

One of the everlasting wonders of the River is that a significant percentage of these pleasure boaters aren't wiped out each summer by the big towboats. Not that the towboatmen are at fault. Once under power, especially if they're heading downstream, they are virtually out of control except for steerage. If the captain of a big tow sees a disabled boat in the channel far ahead, he can only hit the horn, back his engines, and pray that the boat's passengers are good swimmers and have enough sense to abandon their craft in time. There is little he can do to keep from running down such a boat, for his string of barges hold the equivalent of several long freight trains and he may be unable to stop for a mile or more. In midsummer when heavily laden barges are running in the nine-foot (average) channel they may almost rub the riverbed in shallow stretches, and such a juggernaut can grind a sizable cabin cruiser into the bottom of the Mississippi with scarcely a shudder.

It is hardly surprising that towboats wipe out small craft now and then; the wonder is that it doesn't happen more often. Sometimes people die when it does—and sometimes they only wish they had. Several years ago, on Pool 26, a cabin cruiser with two passengers was anchored at night well out of the main channel. Purpose: a postprandial romp of a highly intimate nature. But alas, anchors do not always hold as they should, and the cruiser with its two oblivious passengers, dragging anchor with no running lights, slowly drifted out into the channel where a large towboat ran them down. The cruiser was demolished under a quarter-mile of ponderous steel hulls, but by some miracle the two lovers were thrown clear and managed to swim ashore unhurt. Their problems, however, were not over. Not only had a costly boat been destroyed, but they were both naked as the day they were born and without money or identification of any kind. And although they were indeed married, it didn't happen to be to each other. I've always wondered how the Coast Guard handled that one.

I give the big towboats a very wide berth. The bow wave of

those huge square barges can be formidable, to say nothing of the wake kicked up by a pair of seven-foot propellors ("wheels," in river parlance) and twin Diesels generating 7,500 horsepower. My son Chris and I were coming downriver one day in the freight canoe, with Chris at the motor. A big towboat and string of barges had just passed us and Chris cut across the wake a bit too closely for that day's particular combination of wind, current, bottom con-figuration, tow displacement, and a few other things that combined to form the worst stern wave I've ever seen. I looked up from whatever I was doing to find that the bow of the boat and I were poised over a steep brown slope that led down into a considerable hole in the water. The boat seemed to check, to stand still while the yawning hole moved back beneath it and ingested us. Teen-aged Chris had been seeing some California surfing movies, and as we slid over the brink he yelled happily: "Cowabunga, Pop! Hanging ten!" If I hadn't been so scared I'd have crowned him with the grabhook. From the bottom of that brown abyss all we could see of the big towboat only seventy yards ahead was its radar antenna. For once I was glad we were in a freight canoe and not a square-bowed johnboat, for the high clean sheer of the bow met the opposite water-slope nicely, cut deeply into it almost to the breasthook, and then rose to it without shipping a drop of water. In the same circumstance in a johnboat, with my weight forward as it was, we might have filled the boat.

The Upper Mississippi is generally tamer water than the un-dammed, unchecked Lower River. It is an easier current, slowed as it is by the series of channel dams, and it normally flows at about three miles an hour. There are other times, however. The winter and spring of 1982 and 1983 were wet ones through the Upper Midwest, and for several months all the gates of the big channel dams were raised and it was an open river from the Twin Cities to the Gulf of Mexico and running five miles an hour much of the time.

There is a saying that during times of high water the River is "high in the middle from falling away at the sides." Buffon, the French naturalist, noted the same thing late in the eighteenth century and commented in his *Natural History* that "when a river swells suddenly by the melting of snow, or any other cause, the middle of the stream is sensibly higher than the sides; in one instance, the elevation is said to have been as great as three feet." I can't say I've ever seen *that* much elevation when the River is in spate, but I recall looking cross-river during an extremely sharp rise and noticed a commercial fisherman's johnboat that appeared to be hull-down in the water. I remember thinking at the time "He's heavily loaded to be crossing the River in water like this," but as he came closer I could see that his boat had normal freeboard and was riding high. It was cool, perfectly clear weather with no heat or light distortion to speak of, and I'm sure I was looking at that boat across a half-mile of river that was anything but a perfectly flat water surface.

Speed of current is largely dependent on the configuration of the River. The narrower a section of channel, the higher the current's velocity—a simple venturi effect. On bends of the River, the current will be faster on the outside of the bend—a centrifugal effect. On a perfectly straight stretch of river the current is likely to be fastest toward the center and slowest at the edges—an effect of friction. That same friction is apparent in the main channel as well, where the slowest water is at the bottom and the fastest current is more or less at mid-depth where flow is hindered by neither bottom friction nor wind friction at the surface. The current is most powerful just below the channel dams, of course, where the River is freed of constraint and falls a short distance with the pressure of the higher impoundment behind it. The wildest river water I've ever seen anywhere has been below the open rollers of our Dam 26 in the winter when ice was being sluiced off the pool above—a maelstrom of whitish-brown billows that thundered downstream for hundreds of yards.

The ways and wiles of the River are endlessly varied. I probably run under full power more than I should in places that demand a little more caution. Now and then, during low water, I find a wing dam where one shouldn't be—surprises that have cost me the lower unit of one outboard motor and more propellers than I care to remember. (At $50 per prop, this can lead to penury. Doc Kozicky sagely advises me to buy a $150 stainless steel prop and have done with it, but he can't tell me how to get a $150 outboard-motor prop past my vigilant wife.)

There are strange effects out there on the River in a small boat. Deep holes, shallows, wing dams, rainbow reefs, and channel crossings all affect the surface. A billionweight of water, moving steadily and inexorably toward the distant sea, responds in a multitude of more or less violent ways to the shifting inequalities of the riverbed. Whirlpools and strange boils of water, often many yards across, appear and disappear on the surface. Something violent is happening down below. Running downstream under full power, the johnboat planing high and fast, I come onto a broad, swirling boil many yards across and feel the boat slow and skid as if I had suddenly driven up an abrupt and slippery hill, then breaking out of the turbulence and regaining speed and full control. There are other times, during high water, when strong shearing currents grip my heavy johnboat like a dog worrying a bone. Both high water and low have their special hazards; once, during low water in the dead of summer, I was loafing in slow circles off the point of an island waiting for a big tow to pass by when the displacement wave of the heavily loaded barges lifted me into the shallows and let me down on a large submerged stump where I hung for over an hour. There is no more helpless feeling; you can't get out of the boat for the deep mud, and the waterlogged wood has a tenacious grip.

I'm never on the main river under full power without keeping a sharp eye out for floating debris—especially when the stuff is being fed into the channel from tributaries and backwaters. I shudder at the sight of water-skiers happily hotdogging on high water—remembering the day I came across a massive corner post wrapped with barbed wire, floating almost submerged in mid-river. If ever an object was capable of blemishing a water-skier's suntan, that one was. I fastened to the awful thing with my grabhook and towed it ashore, but the River reclaimed it a week later.

Even ashore, the River must not be taken lightly.

One of the regular duties of the Corps of Engineers is maintenance of the nine-foot navigation channel by dredging areas filling with sand. This is spewed out into the shallows nearby to form big islands and tempting beaches. When freshly dredged it is a treacherous slurry of sand and water. The Engineers post such spoil banks with warning signs, but on hot summer weekends the signs are often ignored. No river sandbar is really stable, although naturally deposited bars come as close as any. But artificial sandbars are entirely counter to the River's way of doing things. They can be fearfully unstable with flocculent margins, patches of quicksand, and other nasty features. A couple of years ago on Pool 26 a teen-aged boy died after stepping on a patch of wet sand and vanishing into a vertical well where the River was boiling up under one of these spoil bars.

Familiarity with the River breeds contempt only with the short-lived. When experienced rivermen take risks, as they all must at one time or another, they are calculated risks with a full knowledge of consequences. My old friend and gunning companion, the late "Arby" Arbuthnot, worked his nets from a johnboat that wasn't a whole lot smaller than your average ocean liner, and I once needled him about using such a big boat. "Tell you what," he drawled. "I love to go out on the River—and I love to come back." Which was Arby's way of saying that the River must be very fond of fools, seeing as how it embraces so many of them.

In the soft summertime the mayflies appear, coming up out of the River to mate and lay their eggs, some living and loving and dying in a single day's span. Soft, fragile, innocuous insects, their long abdomens terminated with several long, delicate filaments, they are about as helpless as any adult insect can be—defenseless, weak-flying and passive, seemingly created only to glut frogs, birds, and fish.

Vast numbers of burrowing mayfly nymphs occur in the silts and oozes of impounded portions of the Upper River where slack-ening currents have dropped sediment loads. There the nymphs live for one, two, or even three years depending on their species, developing into adult "spinners" that come up out of the River in clouds. This mass emergence is not well understood; it may occur at virtually the same time at widely separated points and doesn't appear to be triggered by a particular stage of water fluctuation or air or water temperature.

Mayflies are weak fliers, seldom going far from where they emerged, and since they are usually attracted by lights they may invade river towns in immense swarms. Except for the sheer weight of their numbers, they don't really hurt anything. They can be unpleasant, though, when you walk through a cloud of them—soft, clinging little things that get into your hair, eyes, ears, and mouth. They aren't usually small, either. Most of the ones we see here along the River are *Hexagenia bilineata*, a mayfly whose body is about an inch long. The lighted store windows along the main street of a river town may be covered with these mayflies, all perfectly quiet and passive, all facing one direction with wings folded and cocked high. Sidewalks may be thick with their soft bodies; I have slipped and nearly fallen while walking on them. On some lighted river bridges the masses of newly emerged mayflies may build up so heavily that traffic is stopped until steep bridge approaches can be sanded or even, in some cases, until snowplows can clear the way. Out on the River clouds of mayflies may cover navigation lights and blot them out—a navigation hazard made even worse by the mayflies massed on the pilothouse windows of towboats. A night or two later there may be hardly a mayfly anywhere, but in eddies and quiet backwaters the surface may be covered with the spent bodies of a generation that has paid its biological dues.

Parts of the Upper River also swarm with caddis flies at certain times. They, too, may be pests in the river towns but on a somewhat different level. Stronger fliers than mayflies, the adult caddises are also longer lived and can cause problems much farther from the River. They can be particularly pestiferous in towns located at channel dam sites; unlike mayflies, the caddis fly nymphs develop on rocks and other objects that are silt-free, and are concentrated on and around the dams where rocky rubble is swept clean by strong currents. The channel dam at Keokuk is the oldest on the Upper River, built on the ledges of the Des Moines River Rapids—the first major obstacle met by the old steamboats. As ideal caddis fly habitat, it presents some problems to the town; there have been seasons when caddis fly hatches at Keokuk were so dense that many people living or working near the dam have developed acute allergies to the insects.

For some reason, I've never been greatly troubled by mosquitoes on most of the River that I frequent. There are mosquitoes, to be sure, but they never bother me when I'm in a boat—even in the backwater sloughs. It's another matter when one is ashore in the floodplain forests. About the only time I venture into those jungles is to visit a ripe heron rookery, sneaking through thickets of stinging nettle and poison ivy where the unstirring air is so gray with a whining mist of skeeters that local folks say you can reach out and close your hand and leave a sort of white place hanging in the air. What with one thing and another, visiting hours at a heron nursery are brief.

That is one kind of insect diversion; there are better kinds, such as drifting quietly under the overhang of soft maples and river birches along the bank of a quiet backwater, and putting the fly rod aside to pour a cup of hot coffee from the battered old vacuum bottle. I may change position for a leg-stretch, going forward and lying on the bow deck of the johnboat and looking down into the clear water to see if there's any action. There always is.

An imbalance occurs at the surface of water. Within the water, molecules are drawn to each other from all directions so that the attraction of any one water molecule is neutralized and balanced by those around it. At the surface, however, molecules are attracted on one side only and upward attraction is lacking. A peculiar molecular tension results, a surface tension, in which those imbalanced molecules form an extremely thin, tight membrane. Seen from below through a good diving mask, this surface film appears as a silvery sheet. Tough and elastic, it will support little patches of fine sand under certain conditions. It will even support a steel bar— in the form of a dry, fine needle.

Since nature abhors any unoccupied niche that could conceivably support life, it follows that such an oddball environment as the surface film of water would be put to use by specialized animals. So I lie there on the bow of the boat, sip hot coffee, and ponder the *neuston*, that community of creatures that exists on or under the surface film of the river slough.

Most of the action is on top, in the *supraneuston*, where water striders are skating along the surface film on feet clad with fanlike tufts of extremely fine hairs. Leggy, spiderlike bugs, these hunters have short, grasping forelegs with which the insect seizes prey and holds it while body juices are sucked by the beak. The middle pair of the three sets of legs power the bug over the water; the rear pair do the steering. I'd always figured the water strider was attracted to prey by minute vibrations of the surface film—tremors alien to any made by other striders. Now it appears that it is a matter of light, that the striders are alerted by tiny flashes reflected from the ripples of some luckless alien that's fallen onto the surface film or has penetrated it from below and is struggling to return. To some creatures, the surface film is a lethal ceiling (or floor) that traps them in an alien environment. This can even happen to the water strider that skates so blithely across the film. If a sudden splash happens to submerge the strider, it may be as unable to penetrate the film from below as from above—and soon drowns. Conversely, a water flea that normally lives in the *infraneuston* community on the underside of the surface film may be accidentally thrust up through the film by the same splash that submerged the strider. The tiny water flea can penetrate the tough film and return home in only one way: by molting its outer skin and then slipping through the film back "into" the water.

The water strider is a true bug, order *Hemiptera*, a group that includes a number of aquatic hunters. For a while I don't see any of the others—and then comes a pair of back swimmers. Striders are everywhere, skating on their polished water-floor; at their feet, only a few molecules of water below, the back swimmers are rowing themselves upside down along the polished silver ceiling. They are deadly hunters of small insects and larvae, working each side of the surface film, swimming on their keeled backs with long, oarlike back legs and occasionally poking the tips of their abdomens up through the surface film for air. Like all hemipterans, they have piercing-sucking mouthparts. I once caught several back swimmers in a plankton net and was transferring them to a specimen jar when one nailed me in the thin skin at the base of a finger. Since then, I have been assured by experts that I was probably more surprised than hurt. You could have fooled me.

A few yards away, tearing around in frantic circles, is a small flotilla of whirligig beetles, each as black and shiny as a halved buckshot. They, too, are hunters, tirelessly searching for any small insect that falls from the vegetation overhead and is trapped in the surface film of water. They are beautifully designed for this sort of work. There is an invisible Plimsoll line along the whirligig's side that divides the little beetle into upper and lower hemispheres. The upper half is water repellent, the lower is not, and the beetle floats exactly half-submerged. The large eye on each side of the head is divided into upper and lower halves at the beetle's waterline so that it has a fully functional pair of eyes above the water, another pair below, and the whirligig can see clearly on both sides of the surface film.

All these are professional hunters prowling the quiet surface of the slough. The onslaught on unlucky insects by the slough's water striders, back swimmers, and whirligigs is immense, to say nothing of the toll exacted by ducklings, swallows, frogs, toads, and salamanders. If anything, the world beneath my boat—down out of sight in the rich brown water and bottom oozes amid thickets of coontail and pondweed, through dark forests of lotus, bulrush, and cattail stems—is even more savage. This is the realm of such Goths and Vandals as predaceous diving beetles, water scorpions, and the giant water bugs that can even seize and kill large tadpoles and small fish. Some predaceous diving beetles produce larvae so voracious and predatory that they are called "water tigers," able to lunge at fish fry that are killed by the sicklelike mandibles. Dead-

liest of all, possibly, are the nymphs of the dragonflies. These long-legged, rather spidery naiads are masters of ambush, stalking, and attack. Some cling to a stem, camouflaged with algae and waiting motionless to bushwhack any prey that comes within range of their deadly attack. Other dragonfly nymphs carefully, deliberately, stalk their prey, while some lurk almost completely concealed in the bottom ooze or sand for days at a time until their victim happens by. The lethal stroke of the dragonfly naiad is with a highly modified lower lip, the insect's labium. The extension of this labium is too swift to be seen, and combined with a slight lunge of the body a large naiad can seize prey that passes within an inch of it. The two lateral lobes of this labium clamp tightly on the prey, which is drawn back to the powerful mandibles to be crushed, torn, swallowed, and masticated in the insect's gizzard. The dragonfly naiad's protein, in turn, may be converted to giant water-bug tissue, which may become largemouth bass tissue, which, depending on my own predatory skill, may contribute to my mortal coil at suppertime. Lying there at the bow of the boat and watching some of the relentless savagery below, I am perfectly content to be the physical and intellectual dominant at my end of the food chain. I can't think of many things worse than sharing my habitat with a super-smart, 150-pound dragonfly naiad.

When I was first in this slough back in April, fishing for crappies in the tops of a couple of sunken trees, it was a barren reach of water from which last year's emergent and floating vegetation had been bulldozed away by ice. Four months later, the place was bursting with greenery. Above water and below, the slough appeared to be a thinly diluted salad. The American lotus were in full flower, the huge creamy-yellow blooms standing a couple of feet above the water, their cloying fragrance heavy on the still, warm air. There had been a pre-dawn shower, and raindrops were coalesced on the waxy lotus leaves in standing puddles an inch in diameter, shining silver and looking like blobs of mercury. Our old friend surface tension again.

I leaned back against the tilted motor and let things take their course. There was no one around, no signs of people and their doings. I had the place to myself except for the redwings that were working on the spongy, top-shaped fruiting heads of lotus plants that had shed their petals. Many of the birds appeared to be juveniles, perching cleverly on the lotus stems or atop the seed heads. Mostly, they seemed to be tearing into the seed heads from below. Wood ducks do the same. Well, it's now or never. The seedpods of the American lotus cure out tough and fibrous and their "yorkey nuts" are as hard as bullets. The lotus seedpods and stems make interesting fall decorations along with dried teasel, cattails, and maybe Indian corn, although we've never quite understood the local practice of spraying them with gold paint.

The plant zones in this quiet backwater are classic lowland-emergent-floating-submerged. The long spit of land over there has a grove of silver maples and river birch that tend down into boggy ground and dense stands of four-foot-high *Sagittaria*, or arrowhead, whose roots Indian women once grubbed out with their feet and ground into nutritious flour on their *metates*. Beyond that is deeper water and the first of the floating vegetation, the lotus. Farther out, too deep even for lotus or cow lilies, grow the submerged beds of pickerel weed and coontail. A week earlier, I had poled the boat across this same bay on a cooler, cloudier day and started many fish from the quiet shallows. Some were big—carp and buffalo, probably, although I like to think that a few were trophy bass. Such places as this can be fish heaven in the dog days of a midwestern August. The river water, tinted brown with tannins and infinitely fine suspended sediments and plankton, tends to absorb heat more readily than the colorless waters of a lake or brook, especially in the quiet, shallow backwaters. There is always the potential of overheating, losing most of its oxygen, and becoming a death trap for fish. But then there are the plants—the lifesaving aquatics that recharge the water with oxygen, screen out most of the sun, and host the multitude of organisms that constitute the foundation of the River's food pyramid. Of course, the stuff can be almost impossible to fish. But if you ease up to a bed of lilies or lotus pads, find an open space in the big floating leaves, and manage to drop a bass plug just where it should be, letting it rest quietly until the circles stop spreading, and then giving it a small, smart twitch, not overdoing it—well, something dynamic may develop. The quiet little opening in the lily pads detonates; the battle is joined. I can close my eyes and see it, the eight-pound bass erupting from the edge of the lily pads, her great mouth open to the lure, gill covers rattling with the violence of her shaking, and then the heavy splash back into the water causing lily pads fifty yards away to tremble. That is about as far as I want my imagination to carry me because from then on the dream takes a turn for the worse. Hooking a trophy bass in such a place, which I have done on occasion, is one thing. Actually landing the fish is another. Still, I keep going back and trying, probably because fishing (as Dr. Johnson said of marriage) is the triumph of hope over experience.

It's barely midmorning and the sun is already heavy.

These quiet backwaters, walled with dense floodplain forest that shuts away any breeze that might come wandering down the main River, are not good places to be in the dog days of a particularly hot summer. The open River isn't a whole lot better. Commercial fishermen crack the dawn this time of year and have run their nets, traps, and lines and are back at the fish markets by the time most people are at breakfast. If there is any pleasure traffic on the River through the day it is likely to be air-conditioned houseboats and motor yachts. The River glows like polished bronze, the distant headlands dim through the haze of heat and humidity, and the flood of sunlight seems heavy and turgid. At the big dams, the deckhands, breaking down strings of barges and lashing them back together with big "wire" tensioned with seventy-pound ratchet bars, are as soaked with sweat as if they had fallen into the lock bay.

On such a day not long ago I put in at Lake City, Minnesota, for a quick run to the lower end of Lake Pepin and the Chippewa River. I didn't plan to be gone for more than an hour, taking only a quart canteen and leaving my toolbox behind. And so, at midday with the temperature already past 90 degrees and relative humidity almost as high, I came bombing blithely out of Lake Pepin to the mouth of the Chippewa where the main thread of the Upper Mississippi is little more than a hundred yards wide. The rest of it flows thinly over rock-solid sandbar—which I struck, breaking the shear pin that locks the propeller to the drive shaft. So there I was with no power, no one around, my water almost gone, and no toolbox. Not that there was any real emergency; I was only a few miles above Wabasha and could easily row downstream. But there was considerable towboat traffic and I didn't care to argue right of way on the narrowest part of the Upper Mississippi with big tows in a fast current. The short trip, made in fits and starts through crushing heat, took until mid-afternoon—when I collapsed gratefully in a cool, dim Wabasha oasis to rehydrate and consider the day's instruction.

Autumn

The season doesn't really begin as a calendar date, or at a positioning of sun and moon, but when the scarlet flags appear. Here at the bottom end of the Upper River, that is likely to be late October as the oaks, maples, and sumacs below the river cliffs begin to flare in bursts of scarlet, maroon, and gold.

There is fall color all through the woods and field edges then, but nowhere is it set off to better advantage than along the river cliffs. Few other places seem to burn with so many hot colors at once—perhaps the talus slopes at the feet of the cliffs have the minerals, nutrition, and drainage that promote such displays. Or maybe it's because of the setting. The rich bands of color are often narrow and arranged on steep banks and near-vertical hillsides so that nothing is obscured and everything is held up to view. If all the variables have fallen into place with the proper proportions of rain, sun, frost, and wind, the pale limestone cliffs rise above the Great River Road in billows of color. The old burnished richness of autumn is on the land again and Dycie, my artist wife, shakes her head in despair, noting that it's hard enough for art to imitate nature in such intensity, and downright impossible to convince a city-bred art critic that such colors even exist.

It is another maddening time of year for the outdoorsman—worse than spring, if that's possible. I want to be everywhere at once, doing everything at the same time, drinking autumn to the dregs. By comparison, spring is an easy basking-and-puttering time. But October and November! Fall fishing has come in, some of the finest and best. A man would be a fool not to be working the autumn upstream run of walleyes and sauger, or fly-rodding the tributary creeks for smallmouths—except that there are such distractions as fall mushrooms and fat squirrels and ruffed grouse and woodcock and teal, with the main push of northern waterfowl on its way, and deer season within sight. These are just too many fine diversions to pull and haul at an old boondocker. Why couldn't they be metered through the year a little more evenly? February and August could certainly use some help.

This was on my mind one afternoon in late October, hunting grouse in the rough hill country north of the Root River in southeastern Minnesota, as I came leg-weary and winded out from under the trees and onto a lofty stone ledge overlooking the River. It was a fine natural bench for a tired hunter, three hundred feet above the rolling Mississippi and framed with gnarled cedar trees. For several minutes I was content—and then a half-dozen Canada geese came winging downriver and I felt a flash of envy for anyone down there in a blinded boat with a spread of decoys out. It was only a flash, though, because even in my gluttony I knew there couldn't be any finer use of October than the one I was putting it to, nor a finer place to do it.

Since mid-morning I had been hunting from bottom to top, starting out in wet tag alder runs after woodcock that had made a fool of me as woodcock usually do. After a couple of hours of riotous fun in which I hunted through a back-breaking labyrinth of low alders, never really straightening up, filling my boots with water, and shooting eight times for one woodcock in the bag, I quit the low ground and headed upslope. A few alders persisted on the drier ground, meeting a long, gentle hillside of grazed aspens, and at the edge between alder and popple I flushed the first ruffed grouse of the day. For once everything came together and, as our English friends would put it, I "grassed the bird with the open tube." It was a plump, red-phase grouse, much ruddier than the slaty gray-phase birds of the North Woods, and when the handsome tail was fanned out the unbroken black edge-band marked it as a male.

Nearly an hour passed before I jumped the next bird from a valley side just under the point of a ridge; my shooting wasn't as crisp now, and I needed the second barrel of the little twenty-gauge. Yet, a brace of grouse is a brace and I was feeling very full of myself as I topped out on the ridge, stepped over a broken fence at the edge of an old pasture—and almost on a bird that exploded like a feathered land mine. The grouse bore straight away, full in the open, and I snapped off two late shots without cutting a feather. Well, I thought, at least none of the boys was here to see it. Old Glen Yates would have flashed that satanic grin and snorted: "Keerist, Madseen! Ain't you *never* gonna outgrow the buck aguer?" With mock solemnity Louie Johnson would have said: "You shot just overneath that bird." Doc might not have said anything at the time, he would just have filed it away in some corner of his head to be retrieved on an appropriate occasion.

So here I was, duly humbled and pleasantly tired, lying on a high limestone ledge and refreshing myself, like Antaeus, by renewed contact with Mother Earth. Two rich luxuries had been hoarded for this moment—a candy bar and a pair of dry, fleecy woolen socks. I took the brace of grouse and the timberdoodle from my game pocket, smoothing their feathers and fanning the tails of the grouse in admiration, remembering a time when my old friend Frankie Heidelbauer had stopped to rest in just such a place. Maybe it was here.

"It was on a big flat rock looking out over the River," he told me. "What a fine place that was, and how I wished you'd have been with me that day, Old Coon. Out of curiosity I opened up one of my grouse's crops. It had just eaten some frost-ripened fox grapes. And since me and grouse have so many other tastes in common, I figured 'Why not?' and wiped off those grapes and ate 'em. And you know, they was plumb dee-licious!"

Like Frankie, I have many tastes in common with ruffed grouse. But there are limits. I don't eat secondhand fox grapes. I don't walk back and forth on a log in springtime and drum my wings against my chest, either—though it wouldn't surprise me if Frankie did.

The rich, burnished panorama of the Upper Mississippi Valley lay before me, with my side of the River already in shadow and the russet-and-gold headlands on either side ranked off into the blue hazes of October. The air was cooling now, but the limestone shelf on which I lay still held some afternoon sun and the gnarled, ancient cedars clinging to the rock's edge still breathed sun-warmed fragrance. The three game birds beside me are just the color of October, I thought. They are pieces of October, with all its rich fullness, the ripening of the year. If I feel a twinge, it's not for the ruffed grouse but for me, knowing that they are now part of October Past and are already slipping behind me, and knowing also that there are more russet Octobers behind me than ahead. My training and experience tell me that subtracting those two grouse from the local population is unimportant—a compensatory mortality that the population can easily afford. Nor do I feel any spiritual pangs for the taking of life. I hunted as well that day as I knew how, which is to say that my hunting was done in a way that shamed neither the birds nor me. Further, I had gotten everything I had come for, which was not just a couple of pounds of meat and certainly not any slaking of blood lust, but the ineffable joy of quality hunting in quality environment—which is the richest expression of quality freedom that I know.

There are many uses of outdoor October, and I savor them all. I could drink this ale-golden month to its dregs and never touch a gun. But without hunting, some of the savor would be missing from autumn. Lovely and rich as autumn would still be, a certain condiment would be gone and I think I know what that is. It is seeing grouse and pheasants and quail and mallards at close-hand like any predator might, and seeing how fine-tuned, ingenious, and intricate their responses to predation can be. I might watch game birds and animals at all seasons under a full range of conditions, and yet never know them as I do when I am hunting them well and they are doing their usual fine job of parrying my thrusts. As hunted and hunter, those grouse and I are bonded in ways that we each understand and accept far better than a non-hunter and non-hunted can be expected to. Some of my literary friends who have taken interest in the hunting mystique are inclined to make light of my half-baked, metaphysical way of looking at it. Still, I can think of no better way of explaining that day's grouse hunt than by saying it had closed the magic circle of man, animal, and land as nothing else could have done. Real fishing can do it too, of course. But then, that is just another shape of real hunting.

To my lasting regret, though, something had been lost in my

fall hunting that I have only just begun to find again. For years, the anticipation I once felt for duck hunting was soured by a dread that could not be shaken off. It had to do with getting up long before dawn to go out into the cold wet darkness. Once I was out there in the duck blind watching the sky pale and listening for the whistle of wings, everything would be fine. It was the dark, and cold, and getting started that were bad, and I knew why.

The English winter of 1944 was one of the nastiest in decades. Our 92nd Bomb Group was quartered in ramshackle tin barracks inherited from the Royal Air Force—as were the scab mites and rats that shared our occupancy. The scab mites fed on us, which we resented, and the rats ate the biscuits in our English rations, which we didn't mind at all. Each barrack was heated with two small sheet-iron stoves fueled with coke from which virtually all combustible elements had been extracted before we got it. The fires in those stoves, like those of the human spirit, burned lowest in the small, fretful hours before dawn—and it was then that we would be called to ring our spears against the gates of Rome.

In a sudden flood of harsh, unshaded light the Voice would summon us to a maximum effort mission. There might be a rattle of half-rain, half-sleet against the barrack, and wind shaking the stovepipes. A good morning to hunt ducks, but not to lift bombers from a slick runway with full gas loads, full ammunition loads, and full bomb loads with those touchy RDX fuses.

There was little talking as we dressed. Someone would turn on a radio for the early BBC newscast to hear what deeds our allies had performed during the night. The signature music was an old nursery rhyme set to march tempo:

> There was an old woman went up in a basket
>   Seventeen times as high as the moon,
> And it was so strange that I had to ask it
>   For in her hand she carried a broom.
>
> "Old woman, old woman, old woman," quoth I,
>   "O whither, O whither, O whither so high?"
> "To sweep the cobwebs out of the sky,
>   And I will be with you, by and by."

There is nothing intangible about the fear of cobweb-sweeping. It has a discrete taste, sound, smell, and aspect. It tastes like old copper pennies, and the foul staleness of a pack of Camels smoked between waking and station time. It sounds like the engines of B-17 Flying Fortresses being run up to check manifold pressure; it smells of high-octane gasoline and vomit. Its aspect is a sudden, angry flush of red against the bottom of the overcast over the horizon, followed half a minute later by the dull *whump* of some bomber over at Kimboulton or Molesworth failing on takeoff, and a little later, through rain-streaked Plexiglas℗, the sight of your own marker lights at the end of the runway spreading rapidly apart as your airspeed builds but your overladen bomber still refuses to lift.

Airborne five miles above Germany between the thunder and the sun, fear assumed different qualities: the raw thin taste of oxygen, and the aspect of black crosses on small, drab airplanes. But there would be none of the corrosive, bowel-loosening dread that was felt down below in the pre-dawn blackness and wet—the same conditions that inspire rejoicing among dedicated duck-hunters.

Such foolishness doesn't affect any of my contemporaries. My friend George Arthur was a night fighter pilot in that same ancient war, and rolling out in the darkness today doesn't bother him a bit. It is the same with Frankie Heidelbauer, Charlie Dickey, and Rocky Rohlfing, who were all aviating while being shot at. But then, I've a hunch that those jokers may not even bother going to bed during duck season. They're that kind. Anyway, I'm pleased to report that at last I have almost outgrown the old pre-dawn collywobbles.

High on the stone ledge in the late afternoon sun with a trio of fine game birds beside me and the shadows of the Minnesota headlands advancing across the River toward Wisconsin, the shadows of old battles are grown remote and indistinct. If they must linger, let it be in museums and books that no one wants to read—not here in October. So I pull on my boots, smooth the plumage of the birds and carefully put them into my game pocket, and head off down the hillside into the deep hollows where dusk is beginning to gather, returning to the River, and to ducks.

*   *   *

Of all the forms of hunting in the Upper Mississippi country, none is more traditional, revered, and assiduously practiced than wild-fowling for ducks and geese.

It began as part of the subsistence of early settlers, expanded into commerce and market-hunting, and became regulated and refined as a recreational pursuit. To the uninitiated, it is a curious form of masochism; to its adherents, it is simply a way of life, and a November without duck-hunting is as inconceivable as one without Thanksgiving.

Market-gunning for ducks, geese, swans, and shorebirds got off to a much later start in the Midwest than in the East. There was local trade in big and small game along the Mississippi in the mid-1850s, but markets were limited. The big cities of the East were ready marketplaces served by rail systems that linked them with such commercial fowling grounds as Chesapeake Bay, Currituck Sound, Barnegat Bay, and Chincoteague. By the 1870s, however, the Mississippi had been bridged, and rail systems were proliferating in the trans-Mississippi country. The outlying farm regions of the Upper Midwest were being tied to the population centers of the East, and when reliable refrigerator cars were developed in the 1880s, a vast new market for wild meats was assured. Such commerce was no longer limited to Saint Louis and Saint Paul–Minneapolis; these now became terminal points from which a flood of dressed prairie chickens, passenger pigeon squabs, venison saddles, curlews, ducks, and geese began to flow eastward from rail terminals along the Great River.

All the cards were stacked against game. As the guns, ammunition, access, and sales outlets of the professional hunters were greatly improved, their numbers grew. At the same time, an almost explosive settlement of the land was taking place. Forestlands were shrinking, the virgin prairies were being broken, with bull ditches and the newfangled clay field tiles bleeding away the old marshes and swales. Game was not only being decimated by commercial and subsistence gunning, it was being denied the chance to recoup some of the losses by natural reproduction. Elk and bison were long gone from the Upper Midwest, as were marten and beaver. Passenger pigeons and Eskimo curlews were on their way out. The big flights of waterfowl continued to trade up and down the River in spring and fall, largely because the best of their breeding grounds had not yet been ravaged by plow and cow. But even in the full flush of commercial hunting there were some bleak omens—during the 1890s there were years when migrating waterfowl seemed to have deserted the Upper Mississippi entirely, with hardly a duck or goose to be seen. By the turn of the century, when the Lacey Act outlawed the interstate sale of wild game (in effect, if not in practice), it was clear that the sky was somehow emptier than it had been only twenty years before.

By the late 1920s and early 1930s, wildfowl along the Upper River were no longer being hunted in spring, nor for the market, and hunting methods and kill limits were being steadily tightened. And on the Upper Mississippi River the vital last step—the provision of suitable wildlife habitat—had been successfully promoted by a Chicago businessman named Will Dilg.

One of the founders of the Izaak Walton League in 1922, he was that pioneer conservation group's first president. Dilg loved the Upper River country, especially around Red Wing, Minnesota, hunting and fishing there whenever he had the chance. It was on one of these fishing trips that his young son was drowned in the Mississippi—and soon after that Dilg gave up business and devoted himself wholly to the league and resource conservation.

As a topflight advertising executive, Dilg had always been an idea man with the gift of persuasion, and one of his best ideas was of a vast wildlife refuge along the Upper Mississippi that would be shared by Minnesota, Wisconsin, Iowa, Illinois, and Missouri, with the Department of Interior as resident landlord. A classic example of an idea whose time has come it swiftly gathered momentum, and on June 7, 1924, the Upper Mississippi River Wild Life and Fish Refuge was created by an act of Congress. Today it is something like 194,000 acres of river bottom stretching from the mouth of the Chippewa to Rock Island, Illinois, 284 miles dedicated to wildlife and fisheries resources. For many years this was *the* refuge system on the Upper Mississippi River. Then, in 1958, the Mark Twain National Wildlife Refuge was added—a 250-mile stretch reaching almost to Saint Louis and including 23,500 acres of bottomland timber, wild islands, running sloughs, and backwater lakes. As luck would have it, this vast playground just happened to be ordained the same year I moved in next to it—putting my already belated maturity on indefinite hold.

Chris grew up hearing wild stories about the Armistice Day storm of November 11, 1940, and of the hunters who died on the River between one noon and the next. A couple of years ago he persuaded *Outdoor Life* that the subject might be of interest to that portion of the magazine's constituency devoted to waterfowling, and he headed upriver to dig out a story. The result was a fine article that swelled this paternal bosom with pride and made me thank the red gods that I wasn't hunting that day on the Mississippi. Johnny Cole and I were jumpshooting mallards over on the Skunk River during the height of the storm, and that was bad enough.

It was a clash of two powerful weather systems, a low-pressure trough that came in from the west and veered south and then north, and an arctic air mass that arrived about a month early. These collided over the Upper Mississippi that Armistice Day morning in a deadly mating of high wind and deep cold—a winter hurricane with wind gusts of eighty miles per hour. At La Crosse, Wisconsin, there would be a barometric low of 28.73 inches of mercury, a record that still stands, and in sixteen hours the temperature dropped from 54 degrees to 9 degrees.

That morning, with the full force of the storm yet to strike, it had seemed an ideal time to be out on the River with a good spread of decoys. The sort of sharp weather change that moves ducks was in the making, and if there is ever a time when the main push of waterfowl will be coming down the river it is likely to be toward the middle of November. Best of all, schools and some businesses would be closed for the holiday.

As it turned out, the hunters were right. There was a major flight underway, and the weather change would catch the migrants by surprise and cause them to tumble gratefully into the stools of decoys on sheltered coves and backwaters. That much was the stuff that hunters' dreams are made of. The rest was pure nightmare.

The storm caught the hunters by complete surprise. Some were clad reasonably well for the developing conditions but many were dressed much too lightly. One of the lucky ones was Emmett Flick of La Crosse. As Chris learned, Emmett

got hold of a few friends and headed out to the river when he saw the weather change. The waves were high enough to give them some trouble as they launched their 26-foot inboard commercial fishing rig but, once they got out on the water, the big boat handled the river well. They headed for an island they knew, a place with a shack and some firewood and a few inviting corners for ducks. As it turned out, there was no need for decoys. The mallards poured into the lee of the island, dropping into sheltered coves without hesitation. The hunters only had to wait on the bank to shoot all the greenheads they wanted. The swells were running three to four feet high by the time the party had finished, so they retired to a cabin to wait out the blow.

It was the kind of shooting that makes a duck hunter forget the cut of the wind and, all along the Upper Mississippi, waterfowlers reveled in it—for a while. By the time they appreciated the real ferocity of the storm, many of them were committed to staying where they were. The lucky ones were on islands high enough to be out of reach of the terrible surf. If there was dead timber, they built fires and waited for a break in the weather. Where there was no firewood, they burned their decoys and boats or huddled together against the cold.

Years later, I heard of several commercial fishermen who had been caught on an island that afternoon and stayed alive by burning their tarred nets and fish boxes, and by breaking up and burning their wooden johnboat. I was never able to flesh out that report, but there is little doubt that it was mostly true. If so, they were among the lucky ones that were on high islands above the waves, with a bit of shelter to shield them from that awful wind, and something to burn.

There were others not so lucky. Chris tells of four Saint Paul men who froze to death on Prairie Island near Red Wing. A fifth man, Bror Kronberg, got to shore in a small boat.

He was one of the few men dressed for the weather. He wore a heavy wool shirt, a sheepskin vest, a canvas hunting coat and a knee-length football warm-up jacket. Searchers found the warm-up jacket near his beached boat—coated with more than 30 pounds of ice. They found Kronberg a few hundred feet away in the lee of a haystack where he had apparently reached the end of his endurance. At least two other bodies were found in similar macabre settings. With desperate strength, a few men had managed to fight their way to shore only to find that dry land was no refuge. The storm had swallowed up landmarks, lights and all sounds meant to signal wandering hunters.

Carl Tarras of Winona was out on the river with his two sons, Gerald and Ray, and a friend, Bill Wernecke, when the wind came. Cut off from the mainland, they made their stand in a stretch of shallow water in the scant shelter of a bank of cattails. Wernecke went first, then 16-year-old Ray. Gerald and Carl hung on into the daylight of Tuesday morning, beaten alternately by subfreezing waves and the gale. Carl died just a few minutes before rescuers arrived. Word among rivermen has it that Gerald had dug himself partway into a muskrat house when he was found. Only his back was exposed to the weather. That shelter may well have made the difference.

Some of the survivors managed to fight their way ashore, some endured the storm in tiny scraps of shelter on islands or in river marshes. Ray Sherin of Winona was fourteen years old that day, hunting with his twenty-year-old cousin Bob Stephan and nineteen-year-old Cal Wieczorek. Aground on a low sandbar, they turned their boat over, spread a mat of bulrushes under it, and lay down together covered with Bob's raincoat. With first light they launched the boat over the ice shelf that had built up around the sandbar and let the wind take them, running aground on heavy ice a hundred feet from the main shore where they were rescued by a skiff from the Corps of Engineers launch *Chippewa*. Bob spent a week in the hospital with frozen hands. Ray, who had been thoroughly soaked by the high waves, was badly frostbitten. He was released from the hospital six weeks later, just in time for Christmas, fifty-eight pounds lighter and minus part of one foot. Yet, like most of the rivermen I've met who suffered through that terrible day and night, the ordeal didn't sour him on the River. He still goes out there whenever he can, hunting, fishing, and doing his fine wildlife paintings out of a remote cabin in the Trempealeau Bottoms.

Some rescue efforts were attempted that Armistice Day afternoon, but there was little anyone could do at the height of the storm except stand along the shore trying to see through that gray wall of wind, snow, and spume. Iowa game warden Dan Nichols, who headquartered in Muscatine, managed to get his big twenty-foot johnboat launched to rescue hunters stranded on a low island. Dan had a plywood cabin on that boat, high enough for a man to stand in, with running lights on the roof of the cabin. When Dan pulled away from the boathouse that afternoon and entered the main River, people watching from the boathouse porch ten feet above the dock could not see those running lights as the boat went into the troughs between waves.

A La Crosse man made an interesting remark about those waves. Ray Bice, whose son was hunting some low islands off Brice's Prairie on the Wisconsin side, could do nothing but watch helplessly from the bank and pray for the wind to drop. At dawn on November 12 the snow had let up a bit, but the wind was still high and the cold was deepening. "It was a strange thing to watch the flowage freeze in a wind like that," he told Chris. "Those big rollers slowly jelled. The fragments of ice in the water thickened it into a slush and then the waves just stopped."

The wind began to fall in the afternoon of the second day of the storm; it became clear and calm, with the temperature dropping to around zero. And the body count went on. I've never seen a complete tally of the deaths caused along the River by that single storm. At least sixty-five sailors died on Lake Michigan that first night, and there was a scattering of fatalities due to highway accidents and heart attacks brought on by snow-shoveling. The official toll appears to be 161—of which nearly half were men who died in their boats or their duck blinds or were lost in the blinding storm after having reached shore. The bottom figure appears to be 40, while the maximum toll of hunters on the River during that fearsome afternoon and night may be somewhere around 70.

It has been said that many of those lives would have been saved in this day of weather satellites, computer-enhanced radar, and with trained meteorologists lurking in every television set. But we must also consider the fact that the men and boys out on the River that day weren't your ordinary rational, mentally competent, look-both-ways-before-crossing-a-street citizens. They were duck hunters, and all duck hunters are at least a little flakey. Those living near the River tend to abandon hearth, home, and reason around mid-October and usually regain their sanity barely in time for last-minute Christmas shopping. I shudder to think what would happen today if our highly efficient electronic media warned of a terrible storm swooping down on the Upper Mississippi—a weather system so severe that it was battering hundreds of thousands of fat mallards down out of the sky and forcing them to abandon all caution. It would have about as much deterrent effect as spraying gasoline on open flame. If you really want to keep duck hunters off the River, you tell them it's going to be a balmy, sunshiny 70 degrees with bluebirds around. You don't dare tell them there's a record storm on the way.

In his account, Chris concluded,

There's no explaining the lure of such a day to someone who hasn't experienced it. There is the shooting, of course; mallards that have hung warily out of range are suddenly driven into the decoys like teal. But the main attraction is something far deeper. The memory of storms such as the Armistice Day blizzard lurks just out of sight over the marsh horizon—sullen, indifferent, powerful, a presence that strips three centuries of fences off a settled land and asserts its own primeval will. Generations of waterfowlers have been drawn by that whisper of the old times; and on the Currituck, the Chesapeake, the Great Lakes, and the Mississippi, some of them have died.

Just as old songs remind us of old loves, the smells of the River conjure up special things for me.

The smell of wood smoke in an autumn evening starts me remembering the October dusk when my daughter Kathryn and I were bringing our boat in to the landing after running hoop nets and basket traps in the River near home. We had a good haul of channel catfish and bigmouth buffalo and Katy was at the helm and we were rolling home in high good humor. An egret was flying high above us, its snowy plumage tinted faint salmon by the last of the sun. Up ahead, through the gathering dusk, came a blaze of lights and happy noise—the little excursion steamer *Golden Eagle* bound upriver with the frailing of a banjo and people singing, and someone howdied us from the texas deck and we howdied back and the whole larboard lower deck whooped and waved at us.

We came into the landing at Grafton and there was the smell of wood smoke. Someone was smoking fish, but there was something else in the air, too, the smell of a wood fire under a batch of apple butter, slow-cooking it in a huge iron kettle in the old way that Shirley Ringhausen does it. For years, wood smoke on crisp air had reminded me of evening camps, hunting cabins, supper for famished outdoorsmen, and foxhunters gathered on hilltops listening for the faint far clamor of their hounds striking trail. Now all that was overlaid with a newer memory, of coming home of an autumn evening with Katy at the helm and the landing just ahead, of an egret beating its way down the sun-lanes while the world below was in purple shadow and the little *Golden Eagle* churned happily up the channel, the crisp air seasoned with the tang of burning hickory and applewood.

Summer has its fine River smells, especially at night when blessed coolness pools under the limerock ledges of places like Infidel Hollow and Jug Hollow and then flows down the creekways into the main valley, with the breath of ferns and wet stone and smell of hidden spring-seeps that have never known the sun. There is a special smell to night water and wet sand that goes with tending trotlines from an island camp and the clean fresh smell of fish taken off the lines under the summer stars.

The smell of faraway skunk is one of the best. It really is. It's not at all like skunk-smelled-close, which is a foul, blinding stench like no other. But when that same smell is diluted with distance, borne on autumnal airs, and slightly seasoned with the tang of wood smoke, it is the essence of wildness.

When I was fourteen, Christmas brought me a new Marble's "Ideal" belt knife with five-inch blade, brass hilt, and a handle of oak-leather rings. This was the knife that skinned my first skunk, a reckless piece of work in which I cut too deeply and managed to slice into the scent glands. I wore that scent for weeks, and the knife never did lose it. Years later I carried the knife, honed razor-sharp, strapped to my leg on those visits to Hitler's Germany. In a place and time that were notably lacking in any pluses, all I had to do was press the warm leather grip of that old belt knife to my nostrils and, lo! the faint lingerings of skunk, wood smoke, home, and Christmas.

Granted, there is no accounting for the odd tastes of eccentric outdoorsmen, but some of us revel in effluvia that offend normal people. I guess you could call us cognoscenti of stink. For example, some of my favorite odors have a sulphur base, although that is generally regarded to be the common denominator for most bad smells. The mephitic stench of the skunk is caused by its sulphur content, as is the rotten-egg smell that emanates from the stirred bottom of a river marsh. Our old friend Paul Errington devoted a distinguished career to marshes and marsh doings, and once took the urbane president of large university out to one of his research areas. Later, when asked what he thought of the professor's wild marsh, the great man grimaced and said: "It stinks." Which it did, for him, but it's all in the nose of the beholder. Such a smell tingles my sinuses like the infernal fumes of a Yellowstone geyser basin, and the wild tang of skunk, mink, and fox. So do the cloying scent of blooming lotus, the peppery freshness of watercress, and the banana-oil sweetness of the Hoppe's No. 9 Powder Solvent that we use for cleaning shotguns. And I know that the smells themselves are not the important thing—it is the images they evoke. If I were a great university president with no river memories or backwater hopes, without ken of wild wetlands and their magnificent vitality, then I suppose such places would only stink for me, too.

Through the wine-rich, smoke-blue days of Indian summer and early fall drift a wealth of rich smells—wild pecans from Royal Landing, the clean sweetness of muskrat peltries from Gilead Slough, duck-blind smells of coffee and hot bean soup, of wet wool and wet retrievers beside charcoal heaters, and gun oil and neat's-foot oil and drying leather, with the rich warm kitchen smells of home after a day of wind and water, and the cedary smell of woolen blankets just out of storage. Then sooner or later there comes a day with a special smell we can always count on along the River— the scent of the first real ice-wind.

Ice does have a smell, of course. "Silvery," Dycie says. "It smells sort of silvery—or is that just the way it feels?" However sensed, by smell or touch, there is no mistaking the ice-wind. It is entirely unlike the chilly breezes of late fall; this wind has purpose, and work to do. It has more on its mind than just shaking down the last of the oak leaves and winnowing ripened grasses on the bluff prairies. It begins by edging the backwater sloughs and lakes with delicate prisms that spread outward from the edges of sheltered coves, fixing the quiet waters under clear, smooth panes. Taking strength from this practice, the ice-wind bends its attention to the main River, building stronger prisms out from the shorelines where the currents are slack, working steadily and reinforcing itself with fresh supplies of cold from the north until ice-shelves extend far into the River and the main thread of the channel itself is thick with drifting slush. One morning the slush has congealed into a rough sheet that extends from bank to bank, and the big towboats with their strings or barges plow upriver and open the channel again. For nearly a month the powerful towboats undo the work of Boreas, smashing through the rough-frozen channel as fast as it closes. But the silver breath of the ice-wind always prevails over the vast power of the diesels, and the morning comes when the big tows can no longer break through the channel; they may be fortunate if they are not imprisoned in it. The rumble of giant engines fades into the south, and for two months, and sometimes three, the Upper River is given over to the silver sounds of the ice-wind.

Winter

The Distinguished Research Scientist came booming through the door, his pipe fuming with bargain-counter rough cut and an expression of mock alarm and urgency on his face.

"How can you sit there drinking coffee," said Doc Kozicky, "knowing that those towboats down on the River are fighting an ice gorge? Don't you realize that the poor barge lines are losing money? *Do something!*"

"Well, could I send some money?"

"Of course not!" Doc snorted. "They'd never accept it. Too proud! Get down there with an ice spud or a blow torch or just flick a Bic. Don't you realize all that damned ice is bad for business?"

Nothing inflames Doc's sense of irony more quickly than the federally subsidized towboat industry, and nothing inspires his diabolic sense of fun any more than seeing that industry blocked and baffled by heavy ice from Saint Louis to Minneapolis. For blocked and baffled it was—held helpless by thickening ice that alternately broke free and then froze and broke again with fluctuations in river levels and temperatures, but always seeming to refreeze thicker than before and tending to build up as ice jams in certain narrows and just above the channel dams—rough, jumbled fields of ice cakes that ground and crashed until they again became solidly fixed in place. On the upstream points of some islands the gleaming slabs were piled twenty feet high.

Life goes on, though, even among the jumbled ice cakes. Bald eagles may find fish frozen in the ice and perch there like great carrion crows. There are often small open leads, usually in close to shore, where tight flocks of goldeneyes, buffleheads, and perhaps some canvasbacks and redheads can be found watering and diving for food. I had stopped beside the road one day and was watching some goldeneyes in an open lead just offshore when I heard the frantic barking of several dogs in the distance. I finally found them with the big 10X binoculars; three large mongrels over toward the Missouri shore were trying to run but having heavy going across the rough ice. And just what was *this* all about? There! About three hundred yards ahead of the dogs, six deer. With the white flags of their tails only at half-mast, they didn't seem to be sharing any of the dogs' desperate excitement. The deer, in fact, were plainly in charge. They bounded airily over the tilted ice cakes, easily vaulting obstacles over which the dogs would labor heavily. The deer would stop now and then and look back, sharply alert but not panicked and in full command of the situation. This continued for nearly a mile upriver, the deer never gaining much on the dogs but not seeming to want to—seeming, in fact, to be almost enjoying the whole thing. The dogs finally broke off the chase and trotted wearily back to the Missouri shore while the deer made a leisurely circle and vanished in the thick willows of Scotch Jimmy Island.

An interesting steeplechase and my side won, but it would have been different if that ice had been glass-smooth. Deer are helpless on such stuff and the dogs could easily have killed them all. As it was, I might have gotten into a peck of trouble if the .270 had been in the truck. It would have been a very long shot and there are about ninety laws against firing a high-powered rifle from a public highway over interstate water, but so help me Hannah, I'd have tried for those dogs. I love dogs, having belonged to some dog or other all my life, but deer-killing dogs aren't really dogs. I hate them, as Jack O'Connor used to say, like God hates Saint Louis.

By the time the River's narrows are gorged with grinding, piling ice, it is usually late December or well into January. For weeks the rough, opaque ice has blocked out such sunlight as there is. Submerged aquatic plants in the backwaters are unable to respire and produce oxygen, ice cover has prevented wind from aerating the water, and yet a certain biological demand is being made of the River. Even though all metabolisms have slowed, oxygen is still being used faster than it can be replenished in many areas. The winter fish-kill begins. Gizzard shad are among the first to die; they perish in vast numbers under the ice and are carried downstream through the channel dams and into the open River just beyond. Here, awaiting the dead fish, are bald eagles.

It is no wonder that some people here at the lower end of the Upper Mississippi wonder about the general concern for the eagles. On January 2, 1984, I counted fifty-one eagles perched in trees below Lock and Dam 26; there were nine in one big cottonwood. Now and then one of the great birds would launch itself out over the open water, seeming to know exactly where to go, snatching a dead fish from the surface and easily competing with those other masters of dead-fish-snatching, the scavenger gulls. The harder the winter and the heavier the River ice, the greater the concentrations of our bald eagles. There are periods when up to two hundred eagles can be seen on the Mississippi and lower Illinois within forty miles of where I'm writing this.

They no longer nest here at Pool 26, and our Eagle's Nest Island is only a place name. But in winter they may be so common that they hardly cause comment. I walked from my house down through Hop Hollow to the River one day, tracking a wayward retriever puppy in several inches of new powder snow. It was fine weather, cold and quiet and perfectly clear, a day of alabaster and long blue tree shadows. There was suddenly another shadow, a moving one, crossing my trail. Then another, and another. I looked up to see eight bald eagles through the treetops just above me. Five were adults marked with white heads and tails and dark bodies; the others were brown-and-white juveniles. They milled silently above the trees from which I had started them, little more than fifty yards away but moving off. It isn't every day that you'll flush a covey of bald eagles, and I watched slack-jawed while they vanished beyond the blufftop. Then came the chill of realizing that somewhere nearby, maybe even under the roost itself, there was a plump, ingenuous black Lab puppy that a bald eagle could see in that white snow with its head tucked under its wing. The pup's registered name, however, was Nilo Lucky Streak—and he soon appeared, safe but lonely, gallumphing toward me with happy welcoming noises.

Things along the Upper River slow down in winter, but never come to a standstill. With the first hard freeze of the backwater lakes and sloughs, as the window-pane ice thickens into plate glass just strong enough to bear a man's weight, some rivermen go turtle-hunting. Easing over this thin ice and winter-clear water that is only a couple of feet deep, the hunters watch for snapping turtles that haven't completely buried themselves in the mud. A heavy iron rod sharpened at one end with the other bent into a hook is driven through ice, turtle, and all. The turtles are at their yearly prime, heavy with winter fat, and will bring top dollar at the fish markets. As the ice thickens on deeper bays and backwaters, there are panfish and pike to be hand-lined through holes cut with auger or ice spud. This can produce quantities of the sweetest panfish fillets of the year. Or, depending on your companions, one-of-a-kind haircuts.

Harper's Ferry was enduring its annual January slump when Jim Sherman and I stopped by that time. The Christmas flurry had come and gone, the three-day New Year's party in Jim Williams's Sportsmen's Tavern had finally wound down, fur prices were awful, under-ice seining wasn't producing much, and Three-Finger George

Kaufman hadn't made an interesting arrest for almost a month. "It's this goddam television," he growled. "My most dependable jacklighters are staying home to watch the rassles. Now, ain't *that* a helluva note!"

We were commiserating with him over breakfast at Katie Quillan's Café, fortifying ourselves for whatever the day might bring, when Benny Quillan came in from his barbershop to join us at coffee. Katie's brother operated his one-chair shop in the next room. I never knew if this conformed very closely with the state health code, and never asked, but it certainly was a neat business arrangement. With only one barber there were sometimes a couple of customers waiting—and instead of reading old magazines or listening to Benny lie about fishing they could go back and forth between barber shop and café. I've seen a commercial fisherman finishing a piece of Katie's apple pie while getting his hair cut.

Anyway, Benny came in from the shop with his usual aura of Lucky Tiger Hair Tonic and excitement, talking rapid-fire and outlining the day's Plan as only Benny could—with a great deal of intensity but limited clarity.

"Noon! Goin' up on the Pool. Poke's No-Door Special. You guys, me, Jim, Poke. Red wigglers, minnies, and mousies."

This rendered out as an ice-fishing expedition on Pool 9 above the lock and dam that would employ small angleworms, brassy minnows, and rat-tailed maggots as baits. Poke Adam's No-Door Special turned out to be an ancient Model A Ford from which the doors had been removed to facilitate escape when, not if, the ice broke. Poke had never bothered to register the car. He operated on the premise that motorized craft using the River required licensing only if they were boats, which were floaters by design and function, and which his old Ford was most emphatically not. "Not with no doors, it ain't!" he assured us. "Why, if the ice busts, this old A-Model will go down like a rock. So don't you worry none about us getting pinched for no license."

With these words of comfort we thundered out over the ice with Poke more or less steering. I never knew only four cylinders could make that much noise. "Tore off the muffler comin' back through a stump field after dark!" he screamed over the roaring engine. "Couldn't see. Ain't got no lights!" Some of the time we were going sideways but mostly we went straight on, and after a while Poke let up on the gas and we glided into a large white bay between two islands. Poke cut the ignition while we were still going ten or fifteen miles an hour, turned loose of the wheel, and stepped out on the running board as he watched ahead. It's odd, seeing your driver do things like that. It takes some getting used to.

With ice spuds and axes we cut a circle of small holes in the ice, skimmed off the mush with strainers, checked the depth with a willow pole, set our bobbers to fish just off the bottom, baited up, and went at it. Benny's bait was hardly in the water when his bobber danced and sank and he hauled out a good crappie. "First fish!" he yelled, and Poke produced a brown bottle that he uncorked and placed on the hood of the Special. Benny went over there and stayed a moment and then came back. Then Jim caught a fish, paid a visit to the car, and returned just as Williams and I were heading over that way. What with the good fishing and those visits to the Friendly Creature, we were keeping pretty busy. Whenever the fishing slowed a bit, someone would yell "Move!" and each man would shift to the next fishing hole on his right and maybe it was the walking on the ice or maybe it was the baits moving down through the water, but the change always produced fish. After several of these flurries the original bottle was exhausted and another had replaced it. There were certain intervals, however, when the fishing ground to a halt. We would fall upon barren times, and nothing seemed to revive our flagging fortunes. Then Poke would howl: "Roundup time! Pow-d-e-rr RIVER!" and leap into the Special and tear wildly around the circle of fishing holes, yelling like a banshee as he stamped on the running board with his left foot.

Since then, I have used long plungers to pound air into the water and drive fish toward waiting trammel nets, but never before or since have I seen a stripped-down Model A Ford driven in circles by a half-soused river rat in an effort to stampede bluegills and crappies. Don't knock it. It works. In fact, it worked so well that Three-Finger George later lamented the fact that he'd gone up to New Albin instead of on river patrol where "I could have hung paper all over you violating bastards!"

As it was, we caught fish until we ran out of bait and the second brown bottle was dry, and then we piled into Poke's Ford and roared homeward through the wintry twilight. I don't remember a whole lot about that trip back to town, except for marveling at the wit, warmth, and camaraderie of my good friends. Arrived at Katie's Café, we were awaiting dinner when someone observed I was a mite shaggy about the ears. Generous to a fault, and willing to postpone his prandial pleasure in a friend's behalf, Benny leaped to his feet and volunteered a free haircut. It seemed a fine idea, and we all repaired to the barbershop in the next room where Benny lowered my ears and we relived the day's adventures, with Poke solemnly declaring that the *real* fishing was out on the channel ice just down from the dam in forty feet of water, and that the ice might even be thick enough to drive right out there and fish from the good ol' Special. Think of the walking it would save. I was too absorbed by this and other fantasies to pay much attention to the tonsorial treatment, and then supper was ready, and after that we had to go over to Jim's Sportsmen's Tavern and tell about the fishing, and then it was bedtime.

Sherman and I left the next morning. I dropped him off at his place and went straight home. Dycie met me at the door with a warm hug, then stepped back as I took off my coat and cap. She looked at my hair and gasped "Oh, my God!"

Some time after that, Benny quit barbering and moved over to Waukon where we heard he went into the restaurant business. He still fishes the River, and Big John Spinner says you can hear Benny laugh a quarter-mile away when he catches a fish. Or maybe he's just thinking of that haircut.

Deep winter now, and for more than fifteen hundred miles the Upper Mississippi lies white and silent under the ice-wind. On broad Lake Pepin the ice is nearly three feet thick and still expanding. In the clearest, coldest nights of the year the trees crack and bang as the deep frost drives into their hearts; somewhere far up Pepin a great expansion crack begins, running out of the distance like an express train coming onto a trestle—a long sustained *boo-o-o-OOOMMMM-m-m-m!* amplified by the vast drumhead of ice, passing on and fading as the Great River returns to its slumber.

With first light, the ice-wind is back. It is hardly more than a breeze, an airy zephyr that would be scarcely noticed in spring or fall, but now it carries a chill factor of 50 degrees below zero and strikes into exposed flesh like blue steel.

I walked across one of the river-lakes to visit a place where a commercial fisherman had discarded a seine-load of carpsuckers and gizzard shad. Eagles were said to be there, scavenging the frozen fish. It was rumored that one was a golden, and although it was almost certainly an immature bald eagle, I had to see for myself. The ice was varied, with large patches as smooth as any glass except for the expansion cracks, and adjacent fields of rough crystals that broke noisily under my shoepacs. It almost appeared to be burning as the wind whirled talcum-fine snow over its surface, making the ice seem to fume and smoke. There was no sun or any sign of sun. The sky was a single white blankness that neither promised nor threatened anything. I had left my camera in the truck, for although there was light enough it was flat light without character, providing neither shadow nor highlight. No matter. I never reached that slough with its alleged eagles. The parka hood and thick woolen cap were not enough to turn the edge of that wind. It was literally striking into bone, into the sinuses of face and forehead as a sickening ache that I could feel clear down under my breakfast, and when I put mittened hands over my face there was no sensation in nose or cheekbones. Even in the heavy beaverskin mittens my hands were growing numb—and the eagle slough was still almost a mile away. Enough. I turned tail, the ice-wind scourging me off its River.

Still, I've been defeated enough by river weather to be philosophical about it, and on the way in I considered the fact that this was only forty miles from the place where I had run out of water that August day, drifting on a river of molten bronze and longing for a breath of the ice-wind. Strange that such extremes should have so much in common. Each comes to mind when I am in the presence of the other, and on the arbitrary scale that translates such extremes to degrees of human comfort, the two occasions

were over 150 points apart, yet the ice-wind seemed to burn in much the same way as that fierce sun had burned just six months earlier.

Each of those days was on the cusp of its particular season, as such days usually are, and from then on things would get better. Blazing afternoons always burn themselves out, and sooner or later the ice-wind swings into the South.

Three weeks after I had been driven gasping off the Wisconsin ice, the River was opening at home on Pool 26. Even on sheltered backwaters the ice had become black and rotten; crappie fishermen were looking to their tackle and dreaming aloud instead of just mumbling to themselves as off-season anglers will; the commercial boys were readying their trammel nets, towboat traffic had broken through almost to Dubuque—and the wildfowl were returning.

Even when the backwaters of our part of the River were still icebound, we would see small flocks of Canada geese—vanguards of the thousands that waited downstate in Little Egypt, conscious of the growing duration and intensity of daylight and chafing to be on their way north. No sooner had ice left the Alton Pool than diving ducks began to arrive—the scaup, goldeneyes, and buffleheads that would soon be augmented by redheads and canvasbacks. Pintails were among the first of the puddle ducks, up from the vast Lacassine and Sabine country of southern Louisiana, pulled up the ancient migration routes by clouds of lesser snow geese. Mallards were coming in, too, and in a few weeks the smaller puddlers, the teal and wood ducks, would be arriving. So would the shore birds, the killdeers and sandpipers and the big Caspian terns and delicate least terns, the great blue herons and white herons. On my first crappie expedition of the year, a yellow-crowned night heron stalked to within thirty feet of me and we exchanged unspoken pleasantries—one fisherman to another.

Muskrats and beaver are out again and working; our eagles are gone, but the bull cardinals are hollering from the treetops and Doc Kozicky opines that he'll trade a winter eagle for a spring cardinal any day. From upriver comes word that a near-record walleye has been boated during a late snow squall. Spring, as Horace Walpole once observed, has set in with its usual severity.

We'll take it any way it comes.

5

A GATHERING OF RIVER RATS

The true quality of a dish can be entirely obscured by a keen appetite, of course, and I had a powerful case of hungries at the time, but even with that qualification I believe one of the best meals I've ever eaten was a fish chowder created on an autumn island in the Upper Mississippi by a hard-used old river rat named Charley Gibbs.

We had left his boat landing just after sunup, and the work and savor of the chilly morning had me thinking of lunch not long after

breakfast. By noon, when the last of the skin ice had melted from
sheltered pools off the Butterfly Chute, we had already been on
the River for nearly five hours running and resetting fyke nets and
basket traps. Then we pulled into a quiet slough, secured the
loaded johnboat to a snag and waded ashore to a little clearing in
the willows.

While I rustled dry driftwood, Charley took off his rubber apron,
broke out a blackened dutch oven and a sack of provisions, and

carefully selected a solid, fresh-caught buffalofish from the bin
amidships. With the fire working, he set the kettle in place and
dropped in a half-pound of snowy leaf lard. As this melted he swiftly
"sided" the fish with a thin-bladed, very sharp knife and cubed the
fillets into chunks. The lard was melted, smoking hot. He dropped
in the chunks of fish and turned to his grubsack again, producing
potatoes and onions that he diced into the pot.

With a pothook he moved the dutch oven to the edge of the

fire and we sat on the driftwood log, smoking and talking about the morning's catch and the business of the River. After a while Charley went over to the boat and returned with a water tin, pouring a half-gallon of spring water into the steaming kettle, with a half-pound of butter as an afterthought. He added a handful of mixed salt and pepper, several dashes of Tabasco sauce, and then sat down again to solemnly regard the fire. A lean, tempered riverman with a long and somewhat horsey cast to his face, Charley Gibbs always wore a faintly mournful expression even when things were going fine, even as the fish were beginning their fall run and the chowder was coming done.

He straightened slightly and looked beyond me.

"Hear that?"

Somewhere behind us a bent branch had scraped against canvas. A moment passed, and Gibbs spoke to the willows: "Noisy as you are, how do you ever ketch anybody?"

A short, weathered man in an old hunting coat parted the screen of willows and scowled fiercely.

"Well, what we got here?" said Three-Finger George Kaufman. "A fat young violator and a skinny old violator cooking the evidence, huh? By God, I'll just have to confiscate some of that!"

Gibbs looked into the fire, more mournful than ever. "How come us poor working fishermen always got to feed old game wardens?"

"Maybe it's because they can't catch their own fish," I offered. "You ever known Kaufman to catch a fish?"

"Never have," replied Charley. "Never caught a fish. Never shot a duck. Lives on what he can confiscate, like a gull."

"Damn right!" said Kaufman. "I'm pure predator. Right up there at the point of the food pyramid. I hunt the hunters, and I caught you bastards red-handed. When is that contraband gonna be ready?"

So we sat together and swapped river gossip and all the while the smell from that dutch oven was enough to break a hungry man's heart. Lansing fish chowder, smoking hot, its savory essence riding the autumn wind to bait a hungry old game warden. There was some more ritual cussing, but not even Three-Finger George Kaufman, senior officer of the Fish and Game Division, could scowl and bristle when the chowder was ready on a crisp, blue-and-gold October noon on the Upper River.

My two companions were classics: Charley Gibbs, archetypal Upper Mississippi commercial fisherman, and George Kaufman, the archetypal river warden.

Charley's work world was the labyrinth of running sloughs, chutes, channels, lakes, and pools that lay between Channel Dam 9 near Harper's Ferry and the Iowa-Minnesota border. It is, to be charitable, a mess. I once tried to count the islands in that twenty-six-mile stretch of River and gave up at around three hundred. There are many more than that—some not much more than mudbars and sand shoals, others, like Battle Island, enduring landmarks. Within this maze of islets and islands are the convoluted sloughs that may wander back into hidden ponds or marshes, open up into broad lakes, or simply pinch out on mud flats. Few men know those waters well, and one of them was the late Charles Gibbs, third generation of his family to fish for a living out of Lansing, Iowa.

I first met him about thirty years ago, when he was fifty-three and had been a professional fisherman for over forty years. A quiet man, as men of his kind are likely to be, but not taciturn. A pragmatic man as well, as a River professional must be, and not given to any romantic illusions where his Mississippi was concerned. If he was ever awestruck by any of the sudden revelations of beauty and grandeur that had so often come his way during all those years on the River, he was not likely to bruit it about—which isn't to say he was unaware of such things, even though a lifetime on the Mississippi might make them commonplace. On that particular autumn day he had been in the act of casting his grabhook when he suddenly paused and looked back at me, saying: "I don't s'pose a tree can do much better than that no matter where you go," pointing with his chin at a rock maple on the steep ground above the slough, a tree in its full, vivid, October crimson. It stood by itself with no other maple in view, set off by a pair of brilliantly yellow hickories in a burst of ruby and gold just below the wall of buff limestone. Then the grabhook was cast and dragged for the tail line of a basket trap, the old fisherman again seeming to be oblivious of the autumn pageant that blazed around us.

Like other commercial fishermen of the Upper River, Charley was virtually weatherproof. I have been on the River with him shortly after ice-out when I could scarcely flex my fingers in heavy trappers' gloves, yet Charley worked with bare, wet hands in a March wind that felt like broken glass. The tarred nets were as stiff as wire, and as Charley rolled them into the boat he would casually slap bits of floe ice out of the webbing with his hands, never grimacing nor giving any sign of pain. He wasn't trying to prove anything; he simply worked better with bare hands, that's all, and a man working alone on the River can use all the edge he can manage. You learn to accept the privations. They are the back side of the emoluments.

Charley Gibbs seldom fished in winter, though. He used that time to build new boats and put his nets and traps in good order, making up for it in early spring and late fall when there was floe ice on the River and fish were bunched up. Two-thirds of his annual catch might be made in such spring and fall fishing. At such times he used the trammel net, a double-walled device consisting of a fine-meshed net hanging from the float line with a second net of much larger mesh. Fish striking the fine webbing would drive it right on through the extremely large mesh of the "walling" and be trapped on the other side, where they would hang in a pouch of fine netting. The bin in Charley's biggest johnboat held over a thousand pounds of fish, and there were many days when he had to return to the landing and unload his catch with the nets only half run. Of course, there were other days, too.

During the summer he might have as many as fifty hoop nets and forty basket traps in the River at one time.

A hoop net is a long tube of heavy webbing stretched over metal hoops. The lower, downstream end is open. Fish enter there, pass easily through a pair of webbing funnels, and are trapped in the main chamber. This hoop net lies on the bottom of the river, its upstream end secured to an anchor by a long "tail line." A fyke net is a big hoop net with long fences of netting leading to its mouth. These are staked to the bottom and called "wings."

A basket trap is a wooden version of the hoop net, a cylinder of wooden slats secured by wooden hoops. Such a basket is likely to be seven feet long and perhaps eighteen inches in diameter, with two throats of tapered wooden splines that form cones. Like the webbing throats of hoop nets, they permit fish to enter but prevent them from leaving the trap. Basket traps are used only for catfish and, like hoop nets, lie on the bottom and are anchored by tail lines. They must be thoroughly waterlogged in order to stay submerged, but heavy as such traps are, Gibbs has had them stolen by beavers that cut the tail lines and towed the traps away to build them into their dams.

Placement of nets and traps varies with season, river stage, kinds of fish being sought, and the subtle judgments and instincts of long experience. Charley subscribed to the widely held belief that somewhere in the River are the "fishes' sidewalks"—routes heavily traveled by fish during their migrations and feeding and spawning movements. Such pathways shift and fade with water levels and changes in bottom contours and season and are strongly defined in early spring and late fall, often culminating in winter concentrations that are bonanzas for the fishermen using nets under the ice.

Charley never used buoys, blazed trees along the shoreline, or similar marks to locate his nets and baskets. Such obvious marks invite the idle curiosity of weekend visitors, or even downright thievery—although it is most imprudent to take fish from another man's nets. But although Charley's gear might be sunk in two fathoms of featureless slough that was girt with a blank wall of willows, he rarely failed to pick up a tail line with his first try. He would gauge the shoreline with narrowed eye, line up some mysterious checkpoints, and heave his grabhook—a heavy grapnel with four short, blunt hooks welded to the base of a solid steel cylinder about a foot long and two inches in diameter. Dragged over the bottom, even in deep water and heavy current, it snags the tail line of the fish trap. Up comes the muddy, dripping crate, perhaps with a pounding cargo of fiddlers—those sleek, firm channel catfish from fourteen to sixteen inches long that are the top money fish in the River markets.

When I met him, Charley was no longer putting in the sixteen-hour days he once had, although it wasn't all that unusual to spend

Charlie Gibbs and George Kaufman

twelve work hours on the River when the fishing was good. If he ever regretted his hard, chancy way of life he never gave any sign of it. He could probably have earned a good wage at almost any work he chose; he was intelligent and honest and had a vast capacity for work. A steady job with a barge line would have brought him steady money, kept him on the River, and still left him with plenty of time for fishing. But Charley was one of the old River breed not cut out to punch any man's clock. He did for himself, answering for his own failures and successes. Back there in the late 1950s Charley paid about two hundred dollars each year in various license fees and probably had nearly twenty thousand dollars tied up in nets, boats, motors, and other gear. In a good year he would take in eight thousand dollars—which then had three times the purchasing power of today. But the River is a harsh master whose rewards are seldom commensurate with its demands. Commercial fishing is a dicey calling at best, demanding long hours and unending diligence, for every hour spent ashore means nets unlifted and fish unsold. A hand-to-mouth existence in the truest sense, it has defeated more men than it has sustained. The men who do make it are, in George Kaufman's words, "too windy to drown, too mad to freeze, and too goddam ornery to call any man 'Boss'."

Come to think of it, you could probably say the same things about Kaufman.

I first met Three-Finger George on that notorious political pleasure palace, the Iowa Conservation Commission houseboat.

It is the practice of fish and game agencies to maintain field facilities around their states to house transient personnel and provide meeting places, field labs, and some equipment storage. In Iowa, one of these places was an undistinguished houseboat anchored in a backwater near Lansing and used by game wardens called in to assist with enforcement sweeps, biologists on seasonal fish and game surveys, and for commission meetings. We frequently used the place, and always in line of work. It was certainly no secret. Somewhere along the line, however, the doughty Des Moines *Register* learned of its existence and blew an editorial whistle long and loud. The houseboat was quickly upgraded to Loveboat proportions and, to our considerable amusement, puffed into a *cause célèbre*—Iowa style.

My first visit to this low resort occurred some years before it had been scuttled by the *Register's* exposés. Jim Sherman and I were working not far downriver when word reached us that we were expected at the houseboat for supper, and we reported *aux appétits* a couple of hours before sundown. I was ushered into the galley where a compact, middle-aged person was preparing to fry catfish. Clad in field khakis, T-shirt and white apron, he stood several inches under six feet. His gray hair was combed straight back but tended to stick out on each side, giving him the look of a slightly diabolic night heron. He greeted Jim and gave me a long, searching look as we were introduced. I mumbled the usual courtesies as the cool appraisal went on.

"Well," said Conservation Officer George Kaufman. "Fat little sonofabitch, ain't you?"

Granted, I was carrying a few extra pounds at the time. Still, (1) my waist was eight inches smaller than my chest, (2) my mother happens to be a most human being, and (3) such introductory comment has been known to produce violent exercise. So there I stood, hand extended and wondering if I should throw a punch with it, when the sun burst out of that lined face in a smile as bright and warm as any man ever cast upon another. He seized my hand, clapped me on the shoulder, and roared: "*Goddam* glad to know you, Johnny! Welcome aboard! You make the salad!"

It was the middle finger of his left hand that was missing, and I never did learn how he lost it, but he was "Three-Finger George" until the day he died, ranking with such other legendary game wardens as Owl-Eyes Osborne and Bear-Tracks Rafferty. Nor had that finger been lost in vain.

I was working in the Harper's Ferry area on a magazine assignment with photographer Gerald Massie. Gerry had brought his wife and two children, one of whom was a lively little boy of about six who bit his fingernails. We were all at supper in Katie Quillan's café when George came in from checking net tags on the River, hungry and full of devilment. We invited him to join us, and had finished off some of Katie's apple pie when George noticed the little boy biting his nails.

"Hey! What are you doin', boy?"

"Huh?"

"What are you doin' to your fingers?"

"Nothing."

"The hell you ain't! I know what you're doin'! I used to chew my fingernails, too. And you know why I had to quit?"

"Why?" (timorously)

With a diabolical scowl Kaufman held up the maimed hand from which a finger was missing all the way to the main knuckle. "*That's* why! One day I was chewing on my fingers and they began to drop off!"

We never knew if that broke the kid's habit permanently, but it must have slowed him down for a while. Benny Quillan and I even discussed the possibilities of putting George out for rent to the parents of nail-biters, but dropped the idea when we considered the fact that a kid might stop nail-biting and take up swearing.

I don't know if Three-Finger George could have won a popularity contest along the River. He wasn't the kind of officer who ingratiated himself with anyone, and was the poorest sort of politician, but he had guts and he was fair and was generally regarded as one of the River kind. He once told me there were such things as honest poachers and dishonest poachers, the distinction being one of genuine need—and I rather doubt that any poor families in his territory were ever penalized for having eaten illegal venison, though God help the needy poacher who ever bragged about it in a tavern.

George often used a canoe to quietly work the sloughs and backwaters, and he sometimes left it on the bank of Harper's Slough. One Monday morning it was gone. He was considerably put out about that. So were certain river people, who passed the word downstream and up. A couple of weeks later a commercial fisherman from Sabula spotted the canoe in a backyard, and it was duly returned. We toasted the homecoming in Jim Williams's tavern, with the consensus that the local folks could give Three-Finger George mischief enough and didn't need any help from outlanders down Sabula Way.

When George finally retired he had about fifty years of service with the Conservation Commission. Some of it had been with the old Fish Rescue Crew, I believe, but most of it was served as a game warden. He had no money or property to speak of, and precious little to show for a half-century's dedication to fish and wildlife. But there was a small cabin up in the Yellow River Forest, and the Lands and Waters Division permitted George to stay there as "caretaker." He lived out his life in that cabin, crusty and pepperish to the end, surviving into the Age of Ecology to the discomfiture of earnest young hikers whom he confronted in the back country, telling them that they damn well didn't belong there and bidding them to take their Vibram boots, wine *botas* and candy-assed selves back to the campuses whence they had come, and leave those hills to *real* woodsmen.

This, predictably, caused considerable flap. Strident complaints would ramify all the way to the head office, official apologia ensued, and eventually the placid tenor of the Yellow River Forest was resumed—until the next time it happened, which wasn't usually very long. George Kaufman died curmudgeoning, and with all due respect to the Age of Ecology and its fervent young disciples, something vital and curiously tender died with him.

*   *   *

Today, state conservation officers are likely to be unionized, college-educated, impeccably uniformed, and work a forty-hour week. Old wardens like Kaufman scarcely knew the meaning of schedules; they might not be seen around for a week or more, and then surface unexpectedly at strange times and places.

Iowa Conservation Officer Dan Nichols of Muscatine—the warden who braved the 1940 Armistice Day storm to rescue hunters stranded in Pool 16—was a case in point. I was on assignment in the Fruitland area south of Muscatine one summer when Dan asked if I'd care to join him on patrol.

"Set your alarm for one o'clock," he said.

"On which end? A.M. or P.M.?"

"The one in the middle of the night. Wear long sleeves and hip boots and be ready to walk."

It didn't take very long to spend the easy part of that night. Then Dan materialized at the appointed hour and we drove south to somewhere in the Lake Odessa area, left the car, and took off on foot through the fetid bottomland jungles paralleling the River. To this day I've no idea where we were, or where we ended up. It was a heavy, airless night, slightly overcast and still. Now and then we would be on faint trails but mostly, it seemed, we were plunging through dense brakes of horseweeds and stinging nettles. Several times we waded small, dead creeks, almost hip-deep in clinging gumbo, and then up the steep, greasy cutbanks into heavy belts of willows or floodplain woods strewn with down-timber. Dan carried a flashlight that he used sparingly, not doing me the least good, and he discouraged questions.

We hadn't gone as far as it seemed, of course. We walked less than an hour, and probably no more than a couple of miles or so, but that's far enough in country like that, in the black midwatches of a sultry summer night. Then there was a last belt of willows, and the light of a driftwood fire shining through them, and Dan told me to hang close and keep mum and we broke out onto a narrow sandbar where four men sat and lay about a low-burning fire. There was a huge black coffee pot and several kettles, a grub box at one side, and an old johnboat at the sandbar's edge. A rough, no-nonsense camp with men to match. One of them was sitting on an old automobile seat and was in the act of drinking from a tin mug when Dan stepped into the firelight. The man froze, looking with disbelief over the cup, and said: "Well, now. We got company!"

"Howdy, boys," said Dan. "Anything around here that might interest your old game warden?"

"Where in hell did you come from, Nichols?" asked one of the men lying by the fire. "There ain't a road back that way for better than a mile!"

"Almost two, as a matter of fact," said Dan. "Any coffee in that pot?"

It was a straight camp. At least, as straight as such camps are likely to be. The four men were spending a long weekend on the little sandbar, running trotlines and bank poles, sleeping and eating and yarning. They were all licensed and aboveboard, the fish in their live-box were all legal, and yeah, there was coffee in the pot, and something extra strong to lace it with if we were of a mind. "Much obliged, but just straight coffee," said Dan. "No likker while we're on duty."

"On duty!" said the man on the old car seat. "*On duty?* Man, it's only three friggin' o'clock!"

"Patrol duty never ends," Dan replied. He managed that with a straight face, too. Maybe he really meant it. "Besides," he added cryptically, "there's still an hour or so until first light, and a lot of ground left to cover."

When we got back to the car, sweat-soaked and ravaged by mosquitoes, I asked Dan if he'd acted on some sort of tip. No, he hadn't, except that he knew there would be fishermen camped in that remote place.

"Then why did you even check them out? What was it all about, anyway?" I persisted.

"You might call it public relations," Dan replied. "Nothing like a little midnight visit now and then to keep the boys clean. Couple of days from now they'll have the word spread all over Muscatine about how they were camped 'way to hell-and-gone down by Yankee Chute, and how that crazy game warden came slipping out of the brush at two-thirty in the morning. 'The old bastard never sleeps,' they'll say. Now let's get back to town. I'm gonna sack out."

The freak landform known as Trempealeau Mountain rears up out of the River near the Wisconsin shore a few miles downstream from Winona, Minnesota.

Once part of the rugged headlands that flank the Mississippi in this region, Trempealeau Mountain was severed from the mainland long ago when the River swung around behind it; today it is an isolated monolith rising several hundred feet out of the River and sewn to the mainland by the single thread of the Burlington Northern Railroad, a narrow grade of rock fill that runs along the Wisconsin shoreline, across the base of the mountain, and on northward past the broad marshes of the old Delta Fish and Fur Farm. The mountain itself is singular enough; even stranger is the sight of a long freight train apparently running up the middle of the Mississippi River—often beside an arch-rival towboat.

At the western foot of Trempealeau Mountain runs the Upper Mississippi. On its northern flank is the Delta Fish and Fur Farm and the Trempealeau National Wildlife Refuge. Behind, to the east and south, is the broad, placid enclosure of Trempealeau Bay—part of the natural habitat of a veteran commercial fisherman named Ted Koba.

I launched the big johnboat there on a lowering, rainy morning in mid-May, curious about how Ted was doing out there in the bay, seining rough fish on state contract. In the distance five men were working, laying out net, and I headed directly over to them—striking shallows on the way and ending up by poling my way out of the mud and feeling like a damned fool. No matter; commercial fishermen run aground too, and they were too busy to take any special note of it.

They were making their second seine haul of the morning, having started at about six o'clock, and this would probably not be as good as the first since fish tend to move out into deeper water as the day wears on. I moved in closer, being careful to keep out of the way of the seine haul, and was howdied by a long, lean man who was waiting to stake out the net. Slim has worked with Ted

Ted Koba

for fourteen years and has spent a lifetime on the River fishing and trapping.

"We'll know in a little bit now," he said, watching the long float line of the seine being closed near the boats. "There ain't no doubt that a lot of fish move into places like the bay here at night, especially now in the spring. But they start moving out with first light, mostly. The brighter the day, the sooner they'll do that. Day like this, though, they may stay in here a little longer than usual. We'll see pretty soon."

The float line drew closer in a tightening curve. There were strong swirls just below the surface.

"Buffalo are real leery of getting caught in flat water in the daytime. 'Specially leery, I'd say. They'll come in at night, spawn, and move out in early daylight.

"Sometimes you can come out here at daylight and you can see the mud still hanging in the water where the big fish have moved out. Or sometimes they'll move into a little cut next to a bank where the water's, say, four or five feet deep and not a whole lot wider than that, and they'll just lay in there all day.

"Of course, all our fishing here in the bay is in spring. Come later summer, this place is solid with grass and wild rice and the like. Hell, you wouldn't know there's any water out here then," said Slim.

The thousand-foot seine had been drawn in, its big pocket staked as an enclosure with long iron rods that were driven into the soft, flocculent bottom that is referred to as "mush" in gentle company and simply "loon shit" among the earthier classes of River habitués. The lead line was lifted and the webbing of the seine's bag hung from the tops of the iron rods to form a fish-proof basin of one-inch nylon mesh.

"Remember how some old-timers didn't like nylon netting when it first came out?" I said. "Used to claim it would sing in the current and scare fish. You ever hear that?"

"You hear a lot of things on the River," Slim replied. "Especially from old-timers. Say, I've used plenty of cotton and linen netting and if we still had to use it we'd be bankrupt in a month. Just couldn't compete at all. The dam' stuff didn't last, you see. Seemed like all we did was tar it; we was always tarring it, and nothing would make it last."

A light rain was falling steadily, a veil around Trempealeau Mountain. The five fishermen were wearing heavy yellow or black industrial rain shirts over their black chest waders, working easily and efficiently with little conversation. The pouch of the net boiled with fish of all sizes as the men dipped them out and sorted, throwing select suckers and sheepshead into a separate boat for later eating. Bullheads, crappies, channel catfish and largemouth bass were thrown back; the carp and buffalo were dipped into fish boxes. These weren't particularly big fish; there were several fifteen-pound buffalo, but most of the buffalo and carp probably averaged no more than a couple of pounds.

Ted had waded over to say hello and we watched a large carp being dipped out of the seine.

"Not many of those big ones today," he said. "Did you know that they're buying carp spawn now? It's being processed and salted and put in metal barrels and shipped overseas somewhere. No, I don't know what it's being used for. . . ."

I asked him about those smaller fish.

"No sense in trying to sell any of that stuff," Ted replied. "We'd only get about a dime a pound for them. So I think we'll just stick 'em into a holding pond for a couple of months and then sell 'em for thirty cents a pound!"

Then one of the younger fishermen came over in a small boat and picked up Ted. They headed back toward the entrance of the bay and returned shortly with a big half-submerged live box in tow. The smaller carp and buffalo would be dipped into this and taken to Ted's holding ponds near Lock and Dam 6 where they would be fed ground corn and fattened like feeder steers. A puzzling thing here: when carp and buffalo are held together in such ponds, the buffalo seem to prosper at the expense of the carp. They eat not only their share of the ground corn, but also carp spawn and even excrement as well, and although both carp and buffalo are in prime shape when they are netted from the ponds and trucked to market, the buffalo have often shown the best gains. This isn't what might be expected, considering the carp's general success in competing with the native buffalofish.

Ted had returned full of enthusiasm.

"Looks like there's lots of small fish down by the bridge," he called out, even before he had shut off the motor. "Might be a good idea to make a couple of hauls through there."

The seine's bag was being rapidly dipped empty, and a large softshell turtle appeared in a dip net. It was greeted with alacrity, for its dressed meat would have brought three dollars a pound at the fish market—if the turtle ever got that far. It didn't look as if it would. The turtle was carefully and with significant respect placed in a burlap sack. "Gonna be mighty good barbecued," Ted said to me. "You ever eat barbecued softshell turtle?"

"Never have. Never eaten softshell turtle at all, come to think of it. Snapping turtle, sure. But never softshell. You didn't waste any time getting that one sacked up. Don't want him wandering around loose in the boat, huh?" Ted nodded knowingly and gestured toward the burlap sack.

"They say a turtle won't bite through a sack. That ain't so. I was carrying a big snapper in a sack over my shoulder one time and he took me in the side, right through the sack. Still got the scar to prove it!"

The men were carefully folding and stacking the big seine on the carpeted work deck of a large johnboat. They would be making more seine hauls down by the railroad bridge, probably returning home no sooner than 6:30 that evening. A long day spent laboring in rain with wet netting, deep mud that hampered every movement, dip nets of threshing fish, heavy fish boxes, and cranky motors.

"Sure it's a rough life," Slim said. "A lot of dam' long hours and mighty hard work. Still, a man's his own boss and that's worth a lot. Maybe that's why so many guys keep on farming or fishing even though they know they're being victimized by a whole bunch of middlemen. There's a pretty good number of full-time fishermen out on the River right now; guys who are out of work, so they got to do something. You take Ted, now, he's been at it all his life. Reg'lar old-time professional. Seiner, mostly. Last summer he took out a trotline license and went out just once—and the catfish he took on that one run paid for his license. But he mostly seines. We can use gill nets, trotlines, and seines here in Wisconsin, but trammel nets and basket traps like you guys in Illinois use are illegal up here."

A rough life, indeed. But Ted Koba, seventy-three, seemed inured to it. I commented to Slim that Ted had lost some weight since I'd seen him the previous fall.

"He had to," Slim replied. "Hasn't been too well and has had several operations lately. Him and his brother Jack were both powerful men. Ted was always the heavier, but Jack was tall and broad, too."

With a sigh, Slim picked up a sixty-pound fish box and stacked it easily atop another.

"Reckon there ain't none of us the men we were."

A digression to the matter of equipment:

Commercial fishermen are a self-reliant bunch, by and large, independent as hogs on ice and with their own ideas of what equipment should be. You'll rarely see them wearing the kinds of rainwear stocked at marinas and sporting-goods stores, for example. Such stuff is far too fragile for the use they'd give it. They almost invariably wear industrial-weight oilskins, rubber aprons, and the sort of heavy hip boots called "stove pipes" along the River. Most hip boots have adjustable inner straps meant to be fastened around the upper calves of the legs to secure the boots for walking. I've never known a commercial fisherman who used such straps; the thought of not being able to kick off a pair of hip boots if his boat should sink is horrifying to the typical riverman who, about as likely as not, doesn't know how to swim anyway. But while no one will win any free-style events wearing hip boots, they are not the lethal anchors most river rats believe them to be. I have gone overboard in a strong current of 60-degree water while wearing hip boots and heavy oilskins and managed to surface and swim a short distance back to the boat. I couldn't have swum a hundred yards dressed like that, to be sure, but if it had been necessary I could have shed boots and raingear. Don Edlen, late head of the Sabula fish station, once won a twenty-dollar bet by swimming across the Mississippi wearing hip boots pulled up and snapped to his belt. "Nothing to it," Don told me. "You keep your feet high so that air stays trapped in the boots and then, well, you just take your time and quarter across the current without fighting it."

When I first began going out on the River with any regularity back in the early 1950s, most commercial fishing boats were wooden. Some were made by the fishermen who used them; the majority were built in the little river towns by craftsmen who would add custom features to suit the individual. We've never been able to determine the origin of the term "johnboat." Sometimes it's simply "jo'boat." By any name, the basic design is highly pragmatic but not so simple and crude as it might first appear to be. The old original wooden models were sturdy, carefully crafted boats that might be made of four or five different woods—white oak for certain structural members requiring great toughness and strength, white ash that combined strength and lightness and could be shaped and bent, native elm, perhaps, with clear white pine for seats and decking, and cedar for planking. A well-made wooden johnboat had a slight but graceful rake, was somewhat tapered at bow and stern, and might even have a slight tumble home. It was always square-bowed, for bluffness lends stability to a small boat in which men will be standing and walking, as well as providing a stable forward deck on which to stand and work.

Traditionally, of course, they were rowed, usually by a standing oarsman, since there is not likely to be a middle seat in such a workboat. Years ago one mighty boatman, disgusted at the price of fish in New Albin, Iowa, rowed his three-thousand-pound cargo eleven miles to Lansing only to find that the price was the same there. Rowing such a heavy boat is almost a lost art now. The men who did it were skilled boatmen who often made their own oars from clear, straight-grained white ash. Such oars were seldom used with the metal oar locks so common today, but were equipped with "leathers" and the oars were worked between strong black locust thole pins set deep in the gunwales, enabling the rower to feather his oars on the return stroke.

The first engines used in commercial fishing may have been simple little "one-lunger" gas engines of the kind then used to power washing machines. For the heavier workboats, early automobile engines were sometimes mounted amidships. In some of the biggest boats, an entire automobile—sans wheels and chassis—might be mounted in the boat as a power plant.

Commercial fishing on the River—and sportfishing and pleasure boating as well—were revolutionized by the outboard motor. The early fruits of Ole Evinrude's genius may have been cranky enough, although even then they were marked improvements over any other means of small-craft propulsion on the River. But the modern versions are stunning little marvels of compactness, ruggedness, and efficiency. The amount of steady, dependable power generated by so little engine weight never ceases to awe me. A good outboard motor is one of the wisest long-term investments any outdoorsman can make.

The fact that a modern outboard motor may survive several years of steady use by an Upper Mississippi commercial fisherman is its best commendation. There are, of course, those river rats who are relatively gentle with their boats and motors, but that's not typical. A professional fisherman can not afford to baby his equipment. Income depends on the efficiency with which he can attend the highest number of nets, lines, and traps, and that often means charging full-bore from one set to the next even though the route takes him through stump fields, over closing dams and wing dams, with the little engine's water pump being chewed up in shallows of mud and sand, the propeller bent and broken by stumps, logs, and rocks, or worst of all, the entire lower drive unit of the outboard motor wrecked. Buying a used outboard motor from a commercial fisherman is a bit like buying a used Jeep from a badlands rancher; it isn't the all-time smartest thing a man can do.

On the River today, the 25-horsepower motor is probably the standard minimum for commercial fishing craft of average size, which is to say, a 16-foot aluminum johnboat. Each year, though, the engines seem to get bigger and the average outboard on a 16-foot river johnboat today is probably 35-horsepower. There's a great variation, though. In such quiet waters as Trempealeau Bay, for example, Ted Koba may power even large johnboats with 10-horsepower motors. Speed and the ability to dash out of the way of a big towboat are of less importance in such quiet waters than are the economy and portability of a motor. On the other hand, my old friend Curt Oulson of Grafton, Illinois, has been complaining that his 35-horse motor is just too puny for his big aluminum johnboat.

Curt, by the way, is something of an elitist. His favorite fishing boat is a special custom outfit made to his personal specifications from heavy-gauge aluminum stock. It is an extremely strong, well-made johnboat meant for the heaviest use—and in Big Curt's hands, that's what it gets. Such custom boats are appearing with increasing frequency on the Mississippi because they meet the need for equipment that can hold its own under the most demanding loads and conditions.

My own boat is a 16-foot Polarcraft johnboat powered with a 25-horsepower Evinrude outboard motor. The boat's modest length is a bit deceptive, for it has a six-foot beam and could easily handle a 45-horsepower motor. It has a square bow but the forward part of the hull is a modified V hull that tends to cut into swells and chops far better than the typical flat-bottomed johnboat. One of the reasons few traditional johnboats are seen on Lake Pepin is the light-to-very-rough chop always present on the big river-lake. This calls for a displacement hull that slices into the waves instead of pounding across them, and the typical Lake Pepin sportfishing boat has a V-bottomed hull and a 15-horsepower outboard.

Many commercial fishermen still make nets and traps for their own use, and sometimes for sale. They save money by fashioning their own tackle, and probably produce equipment that's better suited to their conditions and individual fishing methods than any factory product.

Two of my favorite people are Ross DeSherlia and his wife, Marie, of Grafton, Illinois, which is at the junction of the Mississippi and Illinois rivers. Ross makes turtle traps, trotlines, trammel nets, and hoop nets at his home, and sells two models of wooden basket traps made for him across the River in Missouri where they can't be sold because their use there is unlawful. These "baskets" are handsome little devices, skillfully made of slats of red oak. The most effective, Ross believes, are those whose funnellike "throats" are formed of springy, carefully tapered, closely spaced "fingers" of sassafras. The design, spacing, and materials of these fingers are critical. Other makers, such as Ira Smith of Hamilton, Illinois, build their fine basket traps with throats of white oak fingers, but do so because sassafras isn't readily available that far north.

Ross's basket traps come in round and square cross sections. The round traps are twelve inches in diameter and fifty-four inches long, and the last time I checked they cost twenty-five dollars apiece. The square ones are twelve inches in cross section, five feet long, and cost twenty-two fifty each. Ross prefers the round model, and in this relatively small size.

Such handmade equipment often has certain features and individual touches that, to other river people, are as distinctive as fingerprints. Old commercial fishermen have told me that at one time they could identify the owners of every net, fish trap, and trotline for ten miles upstream and down, just by certain characters of design or construction.

Marie DeSherlia is *the* trotline-maker in a family of master trotliners. She once perfected a certain knot that no one else had ever tied, and used that distinctive knot in making her trotlines. One day two men came into Grafton with a heavy catch of fish and asked Marie if she would dress their catch for them. They had not unhooked the fish from the trotline but had simply cut the stagings and left the hooks in place. Marie immediately recognized her distinctive knots and told the men that she'd clean the fish when her husband came home to help. Ross arrived, took one look at the knots, and asked the men where they'd caught the fish. Oh, they replied vaguely, the trotlines were really owned by a friend who had asked them to run the lines for him. "You lie," Ross said. "These are my trotlines. Let's call the sheriff."

"You should have seen it," he laughs. "They left the fish and took off like rabbits!"

At this point, Marie broke in with "Yes, I can make the best trotline on the River—and I could make good trammel nets, too, if father would let me."

"No," responded Ross calmly, "you just don't have the strength in your hands to pull trammel net knots as tight as they ought to be."

"I do, too!" his wife declared.

"Do not," said Ross, and there it lay for the time being. I got the feeling it's been going on for years.

Ross DeSherlia is a handsome man—lean, fit, straight as an oar, with wavy, snow-white hair. At eighty-three he appears to be twenty years younger, and is still doing some active commercial fishing. His most recent health problem has been a bad snapping turtle bite on the outside of his right leg just above the ankle, inflicted a couple of years ago by a big turtle that had gotten loose in his boat. Ross usually puts such turtles into a large can, but he'd forgotten the can that day and had put the turtle into a burlap sack. The wound was slow to heal and left a bad scar. It wasn't the first time; he's noticed a certain numbness in his thumb since another large snapper tore out a sizable chunk of flesh from the inside of the thumb near its base.

Ross was born in the Grafton area and has lived there, or across the Illinois River in the rugged kingdom of Calhoun, all his life. He has never held a regular job—at least, not in the sense that most men think of jobs—although as a youngster during the First World War he worked for a while in a mill that made shovel handles. For seventy years he has been a professional fisherman, trapper, guide, and netmaker. In all that time, except for the short stint in the handle factory, he has never called any man "boss." The River has been his chief, and the requirements for successful employment there have been as exacting as any in industry—skill, knowledge, endurance, courage, and patience. No man can survive a long lifetime on the Mississippi and lower Illinois rivers without

Ross DeSherlia

a strong measure of each. Ross DeSherlia is one of the most successful men we know. Enormously respected in his community, he enjoys robust good health at four-score and three, has a marriage that was made in heaven if ever a marriage was, and can still dance a mean chicken reel. "Of course," he told me not long ago, "you don't plan too far ahead when you're my age. You just take it as it comes, and enjoy life."

Doc Kozicky and I enjoy being with Ross, humble seekers-after-wisdom, as it were. "But have you noticed," Doc says fondly, "how the conversation has a way of running out of substance?"

"I don't know what you mean."

"Well, have you ever asked Ross just *where* it is up in the hills that he finds ginseng? Or exactly *where* he plans to set trotlines next week, and what bait he'll use? And did you notice how Ross answers those questions without really telling you anything?"

"Now I know what you mean."

Such questions, of course, are presumptuous and even discourteous. Ross DeSherlia is not some casual weekend angler to pass information on to fellow sportsmen. He is a professional in a highly competitive enterprise and can hardly be expected to blurt his business plans and secrets to any casual questioner. If he tends to be more close-mouthed than some, it's because he has a good deal to be close-mouthed about.

Ross has the reputation of being a master trotliner, especially for big fish.

"Most of that was done one summer years ago," Ross remembers fondly. "Trotlining for the big ones, I mean. I was running lines that spring and got to noticing that some of the fish I was catching was all tore up. Good-sized fish, too. Channel cats and white perch. I'd hook a trotline and start to raise it and I could feel a great weight on it somewhere, and the blubbers (bubbles) would come pouring up out of the deep water. But whatever that weight was, it'd be lost in the lifting, and I'd have some dead fish on the hooks, all scraped and mangled, some of them four–five pounds or more.

"I knew they were gettin' taken by some real big fish, so I figured I'd go after those big ones.

"You know the Stump Patch? Well, I went up there and caught a big mess of dogfish fingerlings only a few inches long. Brought them home and kept them in a tank and fed them. They grew fast, and by the end of the summer they were eight–ten inches long. There's nothin' tougher on a hook than a dogfish, and I wanted tough, live bait for what I was goin' after.

"I used big hooks, number 6's, and I'd run a hook in the corner of the mouth and out of the eye. You know, I used some of them dogfish several times, unhooking them from a trotline and taking them home and then using them again later on, and they'd be as lively as ever. Nothing tougher than a dogfish. And they sure caught the big ones! Some of my trotlines that summer had no more than five hooks on them, and I might set them in holes eighty feet deep. That's where the big cats were—the big flatheads and blue fultons. I took flatheads up to ninety-seven pounds on trotlines that summer."

"Did it pay all that well?" I asked. "Wouldn't you have been ahead by fishing market-size channel cats?"

"Maybe I'd have been better off," Ross replied. "Maybe not. You have to figure it pretty close when you got six kids at home. Those big catfish were only bringing about a dime a pound at the time, but fish that size add up to a lot of poundage real fast."

Trotlines, of course, are meant for catfish although white perch (sheephead or freshwater drum), dogfish, turtles, gar, and even bass may also take the baits. But the wooden basket traps rarely take anything but catfish. (Although there was a certain rainy morning when Chris and I raised a basket off Six-Mile Island and found a thirty-two-inch eel therein. Transferring that eel from trap to burlap sack, in the rain with wet hands, kept us busy for some twenty minutes. It would have made great Saturday television.)

Such catfish traps may employ a variety of baits, but the most widely used are cheese and soybean cake. The waste cheese is in the form of trimmings gotten from the makers in northern Illinois and southern Wisconsin, sold in great barrels to fishermen up and down the River. This stuff begins where Limburger quits. Nothing, but nothing, could possibly smell any worse, but it's pure catnip to catfish. I have used cheese trimmings in basket traps with success, but can't say that it was much fun. Soybean cake, on the other hand, is clean, convenient, easily stored. It is made from wet soybean meal that is dried in thin layers and then broken up into brownish plates resembling pieces of shattered clay field tiles. It is easy to handle and doesn't stink—at least, not until it has soaked and fermented in warm river water for a while. Then it gets pretty rank. But the finest bait of all, bar none, is a female catfish in full breeding condition. Entice her into a basket trap and your day is made.

Fresh from the maker, a new basket trap is worthless. It must be soaked for at least ten days before use, since fish will not enter a new trap. I commented to Ross that I'd always had the best luck with my worst-looking traps—those muddy, waterlogged, blackened outfits that had not quite begun to rot. He strongly agreed, pointing out that basket traps must be thoroughly river-tempered for best results. But although nearly all my own basket-fishing has been done in summer, some of Ross's most notable success has been in March during high water in certain backwater lakes.

He found that by placing the baskets in muddy water, rather than clear water, prime fiddler catfish were packing into the baskets day after day, so tightly that not another fish could enter. For reasons that Ross has never understood, such tightly packed fish tended to stay alive if taken in muddy water, but died if the water was clearing. During that brief period of high water he took four-teen-hundred pounds of channel catfish in the thirteen- to fourteen-inch size range—surely the best-eating and most marketable size of all.

We've notice, from time to time, that Ross tends to softpedal the hazardous aspects of his trade when Marie is around. She worries about his going out on the River alone at his age, and although that won't keep Ross from doing so, he's careful not to emphasize the old dangers that he knows still lie in wait for him.

We were talking about river boats one day, and Ross's eyes lit up at the memory of the Freeland skiffs that were once made near Grafton.

"Wonderful wooden boats, they were. Up to twenty-four feet long, some of them, kind of narrow by today's style, and usually powered with a one- to four-cylinder inboard engine. Boats you could load with fish until there were only a few inches of freeboard and you could still walk the gunnels and not ship any water. Stable, dependable, heavy-duty outfits."

Marie had left the room to make more coffee. Ross looked out the window at the river filled with floe ice, and added: "I've seen times when I'd be over on the other side and get ready to come back home and find the River was full of ice that had busted loose from upstream. That could happen in just a few hours. You had no choice but to go down with the ice, slowly easing your way sidewise when you could. If you really tried to buck that ice, it would saw a wooden johnboat to pieces in no time." He paused thoughtfully and reflected: "Lord! Some of the chances I've taken out there. . . ."

So far, though, he's always come back. He's brought other people back, too. Ross was the prime organizer of the Grafton Volunteer Emergency Unit—a group responding to mishaps on the River. His role in this wasn't entirely unselfish. Because Ross knows the currents and shifting bottom configurations of his home rivers so well, he has often been called to recover the bodies of drowning victims. Over the years he has recovered twenty-two bodies—a necessary public service, but one that wears on a man. The time came when Ross simply decided the grim task should be spread around a bit more, and the Grafton Emergency Rescue Unit was born.

Not many of these drownings were of seasoned river hands,

although they constantly face the threat of drowning. Last December, two young commercial fishermen from Grafton were crossing Pool 26 after lifting trammel nets over on the Missouri side. They had hit the fish just right, and their big johnboat was loaded to the gunwales with prime winter buffalo. It was well after dark before they started for home. The wind had risen and there was a roughening chop on the open pool; halfway across the River their heavily laden boat shipped water, foundered, and sank. Boat, motor, fish, nets, and all were lost. The fishermen were wisely wearing life preservers and managed to make it to shore, but were in poor shape from hypothermia by the time they were picked up on the Great River Road.

Deckhands on the big tows are sometimes lost, although this seems more likely to occur in and around the locks and dams where the strings of barges are being broken up and reformed, than on the open River itself. Still, it is rough and dangerous work and accounts for its share of dead and maimed rivermen.

Most of the tragedies on the River involve casual weekend recreationists—children, water-skiers, swimmers, careless boaters, and assorted drunks. Dealing with such drownings with any degree of frequency is enough to blight one's soul. Curt Oulson has "only helped recover two bodies, and those were more than enough." One of the great blanks in Curt's comprehension is why people take the River so lightly, going out on it as if it were a backyard wading pool instead of the vast, impersonal force that it really is.

Curt isn't a full-time commercial fisherman. He is a heavy-equipment operator but spends a great deal of time on the River, often setting and running nets before or after a day of construction work. He is a big man, dark-browed and lowering, with the kind of face that you wouldn't backhand in some riverfront saloon. Not that you'll see him in any low dens. Curt doesn't drink. He doesn't even smoke or take coffee. Come to think of it, Curt doesn't have many vices—except for feeding candy bars to Dutch Arnold's black Lab retriever, and scratching his name on friends' new boats.

I dropped by his house the other evening, the last of February. The River was open, more or less, but other than that, spring was still somewhere in Louisiana, and holding. Curt had been floating trammel nets and doing well. He'd been trying a new net with webbing of unusually fine monofilament nylon rather than the usual braided nylon, and on the first float he did poorly. On subsequent floats he began taking fish, and was pleased with the new design. It was a net one hundred yards long, nine feet deep, with walling mesh fourteen inches to the side and a webbing mesh of four inches. No little fish for that net!

He'd been catching buffalo and catfish, with the cats running up to seven pounds, and in beautiful condition. He was excited and downright effusive—for Curt. This is one of his favorite times of year to work trammel nets because the big buffalo and catfish are likely to be bunched; buffalo, in particular, seem to begin grouping late in the year at about the time of the first snowfall, and stay bunched all winter—often in and around brushy shelters of some kind. Come late February these groups begin moving, and Curt will be waiting for them. Carp, he believes, tend to winter back in somewhat deader water, in lakes and isolated sloughs.

When Curt floats a trammel net he stretches it full length in the River, letting it hang from the cork line and drift slowly in the current. He doesn't like the lead line of a floating trammel net to touch bottom. There are fishermen who do, but this is risky; if there are any logs, brush, or other junk along the bottom the net is likely to hang up and must be torn loose. Curt uses a depth-finder to detect major bottom obstacles as well as fish. "I just missed a bunch of big paddlefish today," he told me. "They were right down near the bottom and the net slipped just over the top of them. How'd I know what they were? Well, there were some right up near the surface where I could see their paddles."

Trammel nets may also be staked in place and fish driven to them with long plungers like outsize plumbers' friends that are repeatedly plunged into the water to create a loud, hollow concussive *whoomp* that seems to panic otherwise sensible fish. A net may almost fill with fish and the floats will hardly tremble, although the entire float line of one of Curt's trammel nets was once pulled under by just two midsize carp. I marvel, watching veteran rivermen like Oulson pick a trammel net. It's bad enough when there are only fish in the net, for their struggles have literally sewn the webbing to the wall netting. When sticks of driftwood and cockleburrs and algae are added, the big net becomes a tangled, seemingly hopeless mess. But with some judicious shakes and flips, it always seems to come right.

That morning when Curt came in, prices at the fish market were fifteen cents per pound for carp "in the round," thirty cents for buffalo, and sixty cents for catfish. A few little hackleback (sand) sturgeon were beginning to show up at the fish market, but Curt hadn't been taking any because he hadn't been making any drifts over the sandbars where you would expect to find hacklebacks that time of year. He once took a fifty-four-pound rubbernose sturgeon—the local name for lake sturgeon—but there haven't been any others that large taken near here for several years now.

Curt has a "pocketful of trammel net licenses." He fishes with little else much of the time, and each net must be tagged. Now and then, however, he'll set a basket or trotline.

His biggest trotline catch was a 106-pound flathead catfish taken when he and Arby Arbuthnot were fishing together. The moment they lifted one end of the long line they felt the slow, heavy surge of a very large fish and began working their way along the line, making haste slowly and cutting off hooks as they went, for it is not wise to join battle with a giant catfish while handling a line festooned with strong, sharp hooks. When they finally worked the fish to the surface they could find no way to boat it, for the huge "johnny" was much too large for their dip net. The problem was met with the boat's heavy bowline. As the big catfish surfaced, mouth gaping, the men dropped the line into the open mouth, ran it out through the gills, and secured the monster.

"Mighty impressive," I noted, "except when it comes to the eating."

"I don't go along with that," Curt replied. "Why, some of the best eating you can get is the belly meat of a big johnny."

Out on the River, Curt is a single-minded fishing machine, intense, fiercely competitive, and wholly dedicated to the end of filling his boat as efficiently as possible. Which, of course, is the name of the game. His frequent partner is Rex Paddock of Grafton, a rugged, deeply experienced riverman who can match Curt's pace and anticipate his moves, and is even able to talk him into using such newfangled contraptions as electronic depth-finders.

The three of us were sitting under a shade tree one hot day,

Curt Olson

talking about fish prices, with Rex telling of a time they made a net haul that "dam' near sunk the boat."

"It took the better part of the day just to pick the net after we raised it. Man, we had hit'em! We finally got to the fish market and found out that the owner of the place had heard about our big haul and the buying price of fish had dropped from one hour to the next. We were mad as hell, but what could we do?

"It so happened that a black fish dealer from Saint Louis was just outside. He'd come over to buy some fish from the guy, so we offered him ours.

" 'How much do you want for 'em?' he asked. 'How much have you got?' we said.

" 'I got $75.'

" 'Sold!'

" 'But,' says the Saint Louis fish dealer, 'let's go outside of town to close the deal because I have to buy from this guy here at the market now and then.' So we went out of town and we flat filled up that man's panel truck. I mean, we *flattened* springs!"

"Happiness is a net haul that takes most of the day to pick," I observed, "and a fast boat that can get you to the fish market ahead of the news."

"Yeah, that'd be pretty good," Rex replied. "But you know

what I'd really like to do? Put into the River, say, about a hundred miles upstream with an outfit and my dog, and he's as good a dog as ever growed hair, and just float on down, fishing by day and running coons on the islands at night. A hundred miles of heaven, and I wouldn't sleep a inch of the way!"

Although commercial fishermen may employ almost any kind of tackle depending on season and circumstance, most tend to specialize in certain types of gear. Ted Koba is a master seiner. Ross DeSherlia is a master trotliner. Curt Oulson likes his trammel nets, and up on the Keokuk Pool Ira Smith prefers his beautifully made wooden catfish baskets. He runs some hoop nets, but the baskets are his mainstay.

We were nooning in Ira's fish house near the river bank in Hamilton, Illinois, when I asked him how the catfish were doing. Not good, Ira said. There was no doubt in his mind that the channel catfish population was down; he's still fishing, as he has most of his life, but the catches are growing lighter each year.

"How do you figure it?" I asked. "Siltation, maybe? The Pool filling up with mud?"

"Maybe. Or pollution of some kind or other. But come to think of it, there's all kinds of fingernail clams out there in the River right now and you wouldn't think there would be if the pollution was too bad."

Other fishermen are likely to lay the blame on siltation. Curt blames the incessant heavy barge traffic as much as anything, and many biologists simply believe the commercial catch of catfish is outstripping the River's production capacity. Female channel catfish begin breeding when they are about thirteen to fourteen inches long and just coming into the best marketable size. The greatest impact of commercial harvest hasn't been on the bigger catfish, those, say, in the five- to ten-pound range, but on the youngest adults that never have a chance to breed more than one season, if that. Most of the states along the Upper River are increasing the minimum legal size for commercially caught channel catfish; the legal minimum in Illinois, for example, is thirteen inches and there are efforts to increase that to at least fourteen inches.

Farther upstream, even up into Minnesota and Wisconsin waters, people are singing the catfish blues. John Spinner of Lansing told me: "There's not much doubt about it; channel catfish are dwindling in Pool 9. As a matter of fact, I think they're doing better down in Pools 16, 17, and 18 than they are up here."

"Never thought I'd hear you say that. You've always said this was a real hot stretch of River for channel cats."

"Well, it was. From Dubuque on north it used to be great. But that's changed. No doubt about it. Channel catfish have gone downhill."

"Then how about buffalo?"

"Same thing. I can't figure it. You know, back on the first of March my son Bob and I raised a couple of pond nets that we could hardly lift for the bluegills, bass, northern pike, and such. Oh, there's fish out there, all right! But catfish sure aren't what they were."

As the premium money fish of the Upper River, the channel catfish has a number of problems. Siltation, pollution, habitat changes, and eutrophication of the water are damaging, with overfishing further damaging the fishery in the short run—and there is always the threat that environmental damage could delay or even prevent any recovery from overfishing that stricter regulations might effect. And making matters worse from the commercial fisherman's point of view are the "catfish farms" of the deep South that are producing vast quantities of prime, pond-reared channel catfish for the market. The State of Mississippi alone produced 100 million pounds of dressed catfish in 1982. And even that industry is up against tough competition; each year millions of pounds of dressed, frozen catfish are shipped in from Brazil to be sold well under the price of domestic catfish, either pond-raised or wild. The first such South American imports I ever saw were, of all places, in a Grafton fish market just across the road from Ma Keller's old Blue Goose Tavern where many commercial fishermen hung out. If they needed a reason for their drinking, that was as good as any.

The old times are long gone, and with them the great paddle wheelers and packet boats and the spark-flying, boiler-busting races when cabin boys sat on the safety valves of the boilers while their captains howled for more steam pressure. The upstream rapids have been tamed and the Upper River shackled with its chain of immense channel dams.

But other things are about the same; there is still freight on the River and men to move it. There are still storms, and wrecks, and rivermen lost each year. The deckhands still work hard, play hard, and cuss with inspired skill, as always. In spite of the changes and the new ways, Mark Twain would even today recognize his River and the characters that inhabit it.

Some of their names stick in my mind: "Tiger Red," a red-haired deckhand who was a holy terror when drunk, and "Tangle Eye Red," another redhead with crossed eyes. Or hands like "Gold-tooth Brownie," "Snaketail Kelley," "Paddlefoot" (whose feet, one deckhand told me, "were the biggest living things without teeth"), and "Butterfly Jack," who had a weakness for gaudy tattoos. Butterfly Jack had originally been a sandhog on some of the big under-river tunnels in the East, but quit that line of work and took to the river boats after the highly compressed air on his last tunnel job imploded a half-full pint bottle of whiskey on his hip. "If a pore little bottle of whiskey ain't even safe, what chanct has a man got?" he would ask.

Like other trades followed by men who may be isolated from society and limited to their own elite company for long periods, and whose imaginations aren't hamstrung by parlor niceties, river-boating has produced a rich and colorful jargon. Cooks like "Big Tit Mary" created such appealing viands as "Dead Man's Leg" (a dessert often made of leftovers), "Barge Line Buzzard" (chicken), and "Sawmill Gravy," which was gravy of the lumpier, sorrier sort, resembling sawdust mixed with engine oil. What wasn't eaten was thrown as garbage down the "Dollar Hole" and into the River. Such terms, however, tend to be more colorful than accurate, for the grub aboard big towboats today is the stuff of dreams, lacking nothing whatever in quality, quantity, or variety.

In Mississippi River parlance, a "ramstugenous" person is simply the roughest, hardest, bravest man aboard, "with more guts than Armour's backyard." This is usually a deckhand but in some cases it is the Stud Duck (captain) himself. When a riverman appears to be a mite sickly he "looks like he's all let go" and if he dies he has "slipped his lines." The "Old Folks' Home" is the engine room of a towboat, a "Boogley Boat" is a boat and crew up out of Louisiana's Cajun country, a "Bicycle Boat" is a fast boat with few barges and so is capable of overtaking larger tows, and a "New Orleans Boat" is one bound for the Lower Mississippi. The "Cornfield Navy" is that part of the United States Coast Guard assigned to the River, and a "Please Don't Rain Bag" is a paper box used as a suitcase.

The first riverboatman I ever knew was Eldon "Guttenberg Mike" Vorwald, captain of the old *Minnesota* for Federal Barge Lines out of Saint Louis. "They've always called me 'Mike'," he told me one day as we were locking through Dam Number 12 with ten thousand tons of oyster shell, hay balers, phosphates, and steel, bound for Saint Paul. " 'Mike' was my father's nickname. But there was this Wisconsin captain named 'Mike' so they pegged us with our home towns to clear up the confusion, you see. I've been 'Guttenberg Mike' and he's been 'Prescott Mike' ever since."

Guttenberg Mike was forty-three when I met him nearly thirty years ago. He had gone on the River in 1935 as a twenty-year-old deckhand, advancing to second mate, then first mate, and then serving as pilot for eight years. He had been a captain for ten years when I first knew him. To qualify for his master's papers he was required to know twelve hundred miles of river, including sunken islands, channels, bends, and all the visible and invisible features of the rivers for which he was licensed. On the demand of examiners, he could draw detailed maps of the Upper Mississippi from Saint Louis to Minneapolis, of the Saint Croix from Prescott, Wisconsin, to Stillwater, Minnesota, and of the Illinois River from Grafton to its source at Lake Michigan.

Then there's my friend Edgar Allen Poe, who sometimes has problems with airlines. When making a flight reservation by phone, he has found that airline agents may be reluctant to tie up a seat in the name of a man who's obviously one of your less inspired practical jokers.

It wouldn't be much better if he used the name most of us know him by: Captain "Wamp" Poe. "Wamp" is a River truncation of *wampus cat,* the legendary Kentucky varmint symbolizing pure, triple-distilled, uncorked mayhem. Which, let it be said, is a highly inaccurate appellation. My association with Wamp Poe has led me to regard him as a reasonable, sensible, well-balanced man of business—albeit more colorful than most.

I first met him when he was still on the River, master of the grand new *City of St. Louis* for Nilo Barge Lines. He later ended up in the head office as one of the he-coons of the outfit, a somewhat poignant fate for a man who had spent his life (he was literally brought into his father's wheelhouse as a babe in arms) on the Mississippi and Ohio.

Wamp has a distinction for which many of us would trade our heavenly harps: his great-grandfather figures in Mark Twain's *Life On the Mississippi.* Captain Tom Poe, according to Twain, was the master of a small stern-wheeler on which he lived with his wife. One night the boat struck a snag in Kentucky Bend in the Middle Mississippi near the Missouri boot heel. Captain Poe's wife was in the stateroom of the rapidly sinking boat, and since the stateroom door was nearly under water he chose to rescue her by cutting a hole through the cabin's roof with an axe. However, "the first blow crashed down through the rotten boards and clove her skull."

Twain called it a "strange and tragic accident in the old times." Wamp isn't all that sure it was an accident at all. "From what we've heard, she might have been a little of a hellion," he says. Still, Cap'n Tom Poe's name isn't mentioned by certain relatives.

My son Chris served under Wamp for several summers, working the Ohio River from Paducah to Wheeling, West Virginia, on the *City.* He was eighteen when he first signed on as a deckhand, and at the time I never told his mother of Captain Poe's warning.

"You ought to know about a few things," Wamp had told me. "It's dangerous work, and a lot of deckhands have been maimed out there on the River. Lost feet, hands, arms. To say nothing of the men that go overboard, and maybe they're found and maybe they aren't." But, praise be, Chris came through with flying colors and no serious mishaps, and a head full of Cap'n Wamp Poe stories that I hope he writes some day.

I haven't seen any stats concerning the fatal and nonfatal accidents on towboats and barges on the Mississippi, nor have I made any effort to keep each newspaper clipping that recounts another lost deckhand. But while the towboat industry might tend to minimize the hazards, there is little doubt that the work can permanently impair one's well-being. The barge lines are safety-conscious, to be sure, and any deckhand caught out on the tow without a life preserver may be summarily fired. Still, accidents will happen to men and equipment—and some of them are awesome.

A typical river barge is two hundred feet long and fifty feet wide, made of heavy steel, and capable of carrying hundreds of tons of cargo. If a tow is improperly made up, or if some unusual strain begins snapping the heavy steel cables (called "wires") that lash the barges together, the tow may break up and the barges run wild. In one thirteen-month period recently, there were five accidents along the Saint Louis riverfront in which a total of 130 barges broke loose and caused millions of dollars in damages to shipping and docks. One of the worst was on April 2, 1983, when a tow of four oil-filled barges crashed into the Poplar Street Bridge. A burning two-mile oil slick spread downriver, reached a Monsanto Company loading terminal and a Pillsbury grain terminal, and cost some nine million dollars in damages and cleanup.

There is a sort of primordial energy in the dumb, undirected power of a wild barge—an inexorable juggernaut that is all the more fearsome for moving in such slow motion. A couple of years ago a towboat was heading down the Illinois River just above its junction with the Mississippi when an empty barge broke out of the tow while rounding a bend. It was during a period of high water and there was no steep bank to check the barge, which just coasted on inland. It plowed a swath through the flooded timber for three hundred yards, snapping off cottonwoods, river birches, and silver maples and swathing them like cornstalks, narrowly missing a small bridge on Illinois Route 100, climbing up over the road grade and finally stopping there, blocking the highway. On the far side of the highway, with the last vestige of its momentum, it had casually snapped off an eighteen-inch maple tree ten feet above the ground.

"Yeah, and there's trees a lot bigger than that busted off down there," the proprietor of the café in the little river town of Pearl told me. "You can't see 'em now because of the barge backing and filling over it all. Would you believe that barge just coasted in there without really being pushed or anything? Leastways, that's what the captain said."

"Yeah," one of his customers added, "and you know what the Cornfield Navy said to him? They said, 'You may be a cap'n now, but when you get that barge off that road, you're gonna be a private!' "

"Must have been a real scary thing to see," the café owner said. "Especially for that truck driver that was right there when it happened. Yeah! You hear about that? Just as that barge come sliding up across the highway a loaded semi was going south. Driver hit his air brakes and locked all wheels. Slid across the bridge and just sort of nudged that big steel barge. Driver told me that the sight of that barge coming up over the highway was the damnedest thing he *ever* saw, and he wasn't in no hurry to ever see nothin' like it again! I asked him if it did much damage to his rig and he said 'No, I was almost stopped. Put a few dings in the front end, is all. Now I'm wondering how in hell I'm gonna explain it to the insurance company!'"

Every now and then some renegade barges will put a bridge out of commission, or wipe out part of some waterfront. The wonder is that it doesn't happen far more often than it does, considering the eccentricities of river currents, channels, and the gargantuan proportions of the big tows.

With a full cargo tow, Guttenberg Mike's *Minnesota* was several hundred feet longer than the largest battleship ever built—and on some tight river bends a passenger could almost step out on the riverbank from the boat's fantail as it "crowded the land" to make a turn. The boat itself was 165 feet long, 42 feet in the beam, drew just over seven feet of water, and there were places in the nine-foot navigation channel where "she rubs a little bit," Guttenberg Mike said. It takes some doing to ease a string of barges 100 feet wide into the entrance of a lock that's 110 feet wide. At night, in foul weather, it's enough to give a skipper a case of the horrors. Still, it can be even worse out on the open River. One of the wickedest places on the Upper Mississippi, Mike felt, was in the Davenport, Iowa, area in the narrow chute that was once rapids and shallows.

"Farther up in Minnesota the River is narrow, with lots of sharp bends," he told me. "But it's mostly sand bottom. In Iowa, the channel from Linwood to LeClaire is only two hundred feet wide in some places—just a trough cut through rock. You don't run aground there."

In such a rock channel the quarter-inch steel plate of the cargo barges can be easily ripped open, and valuable sunken cargoes such as phosphates can be salvaged only by dynamiting. This precarious part of the channel is clearly marked but there are places in the River where wind can grip the sides of the big barges and blow an entire tow right out of the channel. Steel cables may be snapped and the tow broken up and driven aground. Wind is most serious in late fall in such areas as Pool 13 above Clinton—a lake four miles wide and twelve miles long—and on Lake Pepin in Minnesota. On such pools Guttenberg Mike has seen waves ten feet high sweep over the decks of a towboat and wash through the lower cabins.

Captain (Gutenburg Mike) Eldon Vorwald

You'll sometimes hear rivermen argue about the relative dangers and difficulties presented by the various rivers. Old Cap'n Bill Heckman of Hermann, Missouri, used to say that the men were separated from the boys just above Saint Louis—with the men going on up the Missouri River and the boys staying on the Mississippi. It's all a matter of perspective. Towboatmen have told me that the Upper River is the worst—narrow, shallow, with small lock bays and twenty-six dams. Some Ohio River men claim the worst, but seem to do so with little real conviction. All these may say that the Lower Mississippi towboatmen are the luckiest of all, with lots of water and no dams. However, there is also heavier current in that undammed Mississippi; I have watched a big string of barges, powered by two towboats, take forty-five minutes to struggle upstream around the big bend above Greenville, Mississippi. And while the Lower River may be deeper and heavier—so are the tows. A big tow on the Lower River may consist of forty barges, and Captain Wamp Poe once took a string of fifty-six barges in one tow—over ten acres of cargo-laden steel hulls.

And again—'way for a lightning riverman.

Ken Lubinski was not born to the River, which is usually the case with your classic river rat. He hails from the south side of Chicago and arrived on the Mississippi via various universities, a Ph.D. in aquatic toxicology, and as a staff member of the Illinois Natural History Survey.

For several years he has worked out of the little survey laboratory in Grafton, attending to parts of the Mississippi and Illinois rivers and examining the effects of barge traffic and resultant turbidity on various river organisms. He is a powerfully built man in his mid-thirties, a handy qualification for an aquatic biologist working with heavy dredging and sampling equipment. He is also a qualified diver, and is one of only a few men who have seen the deep and hidden heart of the River.

There is really only one time that this can be done—in the depths of winter when the watershed is locked into place by ice and deep cold, and the silt-free tributaries run low and clear. In the River itself macroscopic life is held in abeyance; the rich blooms of phytoplankton and zooplankton that help cloud the Upper Mississippi during the warm months are quiescent. The Upper River is slumbering beneath thick ice that restrains barge traffic, and, with no powerful towboats to churn its silts and sands into constant suspension, the winter Mississippi is at its annual clearest.

The Corps of Engineers has a perennial problem: the disposal of sand dredged out of the channel in the never-ending effort to maintain navigation. Pumping the sand out into peripheral shoal areas is one solution—except that many of those edges happen to be useful habitats for fish, shellfish, and wildlife. Another possibility is pumping that sand over into other deep spots of the "thalweg"— the deepest part of the River's channel. However, practically nothing has been known of the thalweg's use by fish and shellfish. Just what was down there? Could it be critical habitat that would be badly damaged by dredge disposal? Enter Dr. Lubinski, in a study funded by the Corps of Engineers and the United States Fish and Wildlife Service.

The little team of technicians headed by Lubinski spent January 1984 at five locations in Pool 13 not far below Bellevue, Iowa. Diving with Lubinski was Robert Anderson of Great Lakes Diving Consultants, Traverse City, Michigan. Anderson was a natural for the job; he is not only a professional diver, but holds an M.S. degree in aquatic biology.

It was cruel work. From air temperatures as low as 15 degrees F., working by turn, the two divers entered the River through four-by-five-foot holes cut with a chain saw in fifteen inches of ice. River water tends to be colder under ice than that of a lake or pond, and the men were working in a Mississippi that was never much more than one or two degrees above freezing. They were clad in "dry suits" worn over two layers of woolen underwear, with heavy cotton coveralls worn over the diving suit, going into the long, narrow thalweg of the Mississippi where some scour holes are fifty-five feet deep.

From each of the five stations a downstream baseline was established. The diver would work along this line, making observations at recorded distances on each side of it and phoning in those observations. He was surface-supported by an "umbilical cord" that carried air from large-volume tanks on the ice above, and which also incorporated a lifeline and an electrical line for the 600-watt lamp by which they worked in that dim brown world of the River's thalweg. With that powerful light, visibility at the bottom of the channel extended as much as four feet. Not very impressive by Silver Springs standards, perhaps, but in a river where the summer diver cannot see his hand against his face mask that is an astonishing degree of clarity. It was enough to permit still photography and the use of a video camera that was linked to a monitoring screen in the portable shed up on the surface of the ice—and for those of us accustomed to opaque midwestern rivers, that's not bad.

One of the things the divers noticed, to their considerable interest and surprise, was that the floor of the River's deepest channel was not a uniform, well-scoured habitat. It varied, with stretches of naked bedrock, then hard-packed sand fluted with ripple marks, and, most interesting of all, tracts of assorted rocks and cobble that were not clean-scoured at all, but coated with algal material that was alive with insect larvae and zooplankton. Much of the thalweg's floor was a rich life zone confirming suspicions that these deep sections of main channel are productive mussel sites and important wintering areas for fish. There were heavy concentrations of mussels at several of the sites, and the divers saw and photographed flathead catfish, channel catfish, log perch, silver lamprey, shovelnose sturgeon and silver chubs. Lubinski and Anderson saw no scaled game fish such as walleyes or bass, not because the deep channel is necessarily devoid of such species but because those are relatively active types in winter water and probably moved away before they could be tallied. Catfish, however, are inclined to be sluggish in cold water and were easily approached. Most sought any shelter that would help break the steady press of current. Catfish were "stacked" in certain deep log piles along the borders of the main channel; others lay behind rocks, and some smaller catfish even nosed in between the empty, gaping valves of dead mussels that sheltered their heads but little else. Most of these wintering catfish showed no gill movement, but with mouths slightly agape and gill covers slightly flared, the flow of the current provided the water movement necessary to respiration. One of the most interesting sightings was a mid-sized shovelnose sturgeon comfortably bedded down and partially buried in a patch of sand.

An instructive adventure, revealing the significant fact that the channel deeps of one portion of the Upper Mississippi are not the relatively barren, scoured, sand-paved troughs they were thought to be. Instead, portions of main channel floor are rich life zones of importance at all times—and of critical importance during winter when thickening ice may close peripheral habitats to much of the River's higher life. John Pitlo, director of the Bellevue Fisheries Research Station, feels that anything that disrupts such crucial winter habitat could be devastating. All-winter commercial navigation on the River, with heavy shipping use of a narrow channel that is virtually the only refuge for wintering fish, is a particularly serious threat—and the indiscriminate pumping of sand from the sailing line into adjacent scour holes of the thalweg is a grim prospect.

"This study is of crucial importance," Pitlo said recently. "We're going to live or die by this study. Either we're going to have sand pumping or not."

It all depends, obviously, on how the sand is pumped, and to where. Ken Lubinski feels there may be some environmental benefits in keeping the sand in the channel and letting the River take care of it—"as long as we can make sure there aren't any fish or other organisms using those areas and that the sand won't be redeposited downstream in other valuable habitats."

Will the studies be continued? As of now, there are no plans to do so. This research was only meant to be a one-winter shot—but it is the kind of dramatic and revealing breakthrough that will not be allowed to languish alone—not if I know anything about the aquatic biologists whose prime habitat is the Upper Mississippi.

Anyway, I told Ken Lubinski that I was composing this chapter on river rats and meant to include him, for although he was overqualified in some respects and underqualified in others he averaged out O.K. He liked the idea. Like my old friend and partner Dr. Ed Kozicky, Ken has managed to overcome the handicap of his long exposure to campuses and has grown, well, *riverish*. Which is as much as you can expect of any good field man.

When Bill Allen, now a field representative for Ducks Unlimited in Minnesota, was with the Illinois Department of Conservation, one of his favorite mini-vacations was with his wife on an island in the Mississippi near Hannibal.

"On the point of that island was a little fish camp maintained by an old-time commercial fisherman named Roy Perkins," Bill

The Delta Queen

told me one day. "He was well into his seventies, a very pleasant and congenial gentleman. He didn't own a car, so he would take a taxicab from his house to the boat landing and he hoped to catch enough fish to pay the cabfare. We were talking one day about reciprocity of fishing licenses on the Mississippi, and he said: 'No one owns this old river. Not really. It belongs to everyone.'"

In turning this account over in my mind, it has struck me that

although the River indeed belongs to everyone, it seems to most particularly belong to old river rats like Roy Perkins. I couldn't begin to list all the septuagenarians and octogenarians I have met along the River over the years—old game wardens, commercial fishermen, towboatmen, ancient hunters and trappers and guides. Whatever their calling, they all shared a common and consuming interest in things riverine—whether the ducks were moving, who was having the best catches of fish or fur, what towboat had run its barges aground, who was finding mushrooms out on the islands, what poachers had been brought to the bar, and whether the River was rising or falling. I often meet such men out there. They are usually solitary and are generally cast in one of two molds: the lean, stringy old man with a shock of white hair, or the burly, balding, high-bellied old man who, like his hands, is broad and thick from a long lifetime of heavy labor on the Upper River. They are absorbed in whatever tasks occupy them, minding their own business and keeping their own counsel, reserved and reticent beyond exchanging howdies, and able to appraise you and your outfit with one swift, piercing look. If they approve of what they see, they may pass the time of day and share a bit of news. I almost never meet one that is sour and bitter; that kind is more likely to be found ashore, probably well inland, out of sight and sound of the River, with no interests but his own aches and pains. Not so with the bonafide, lifelong, incurable river rat. The Ted Kobas and Ross DeSherlias seem to go on forever, still actively courting the River into their seventies, eighties, and even, in some cases, their nineties. What accounts for such vigor, this long-lingering boy-hood? Regular exercise? Fresh air? The simple life and plain food? All those, surely, but in most instances there appears to be a sustenance of spirit and a perpetuation of what Rachel Carson called "the sense of wonder."

Charles Edward Russell, author of *A-Rafting on the Mississip'*, put it another way. He believed that it was a gift of the River with which she (typical old riverman's usage, applying "she" to the Father of Waters) rewarded her devotees.

"One and all having the love of her in his heart, to the end of his days had her for a vital interest," Russell wrote. "Life never grew stale or weary for him. To watch the river and talk about the river and recall old scenes and old stories about the river and snap up every line of news about the river and tell again his days and ways of love for the river—it was enough. Men wear out first in their spirits. No riverman's spirit flagged so long as he could remember the river."

6

DEUX EX
MACHINA

Through the Roaring Twenties, the Upper Mississippi scarcely hummed.

There were some doings along the riverbank, to be sure, where many towns and cities blithely ignored Prohibition and the immortal Bix Beiderbecke was emerging from the river town of Davenport to herald the Jazz Age with his clarion cornet. But the River itself was occupied by only a few pleasure boaters and commercial fishermen, duck hunters, and anglers. The commercial river traffic of the nineteenth century had been sunk by phenomenal advances in transport. The main lines, trunks, and spurs of railroads were everywhere, wonderfully efficient transportation for produce, goods, and travelers. And as if the dominance of King Railroad weren't enough, the first paved highways were beginning to appear in the heartland, and some mail and light freight was even being shipped by air!

The fact was that the river freight just couldn't compete economically with rail freight. Ton for ton, freight rates by water may have been only 80 percent that of rail rates, but the railroads could handle far greater tonnages than the relatively small barges used on the Upper Mississippi. The economics of the day demanded the sort of large-scale shipping capacity offered by a hundred-car freight train, and there was nothing on the Upper Mississippi of the 1920s that could possibly match that. The six-foot channel that had been authorized by Congress in 1907 had little effect in opening the Upper River to heavy shipping. Even if such a channel could have been established and maintained, the plan was devised for such obsolete traffic as log rafts and packet boats. And as it turned out, that theoretical six-foot channel was often much less in the low-river stages of midsummer when no more than four feet of navigation channel could be guaranteed—and towboats and barges with less than four feet of draft are no match for hundred-car freight trains. Any freight coming up from the Lower Mississippi had to be lightered at Saint Louis into shallow-draft barges that could be used in as little as four feet of water, or even less. For freight bound downriver from Minneapolis–Saint Paul, the opposite applied. Such transshipments were inordinately expensive, and so the freight-handling of the Upper Midwest became the virtual monopoly of the railroads. Once-bustling waterfronts of the river towns grew to weeds, and the wharves and docks damaged by flood and ice floes were never repaired or replaced. In 1930 only 527,000 tons of freight were shipped on the Upper Mississippi, and towns like Reads Landing, Minnesota—which boasted seventeen hotels when it was the greatest wheat-loading port in the nation—became virtual ghost towns. If the Upper Mississippi was to have any real future in modern American transportation, it was obvious that something had to be done. And in a notably low-water year in Upper River commerce, the United States Congress did it with the Rivers and Harbors Act of 1930.

A dependable nine-foot navigation channel was to be created and maintained by the United States Army Corps of Engineers from just above Saint Louis to Minneapolis. This would require the construction of two dozen low-water dams in addition to the structures already in place: the dam at Keokuk and the lock at Moline that bypassed the shoals of Rock Island. The new dams would create a staircase of river lakes that rose 335 feet in the 662 miles from Alton, Illinois, to Minneapolis. They were to be intended solely for the maintenance of navigation, and not for flood control or power generation. And, since they were low-profile structures, their impoundments would not threaten riparian settlements or the levee systems that existed below Muscatine, Iowa. In the planning stage the total cost of this vast project had been estimated at $140 million—which turned out to be about $30 million short of the final figure. By the end of 1940 the canalization of the Upper Mississippi was about 87 percent complete, and river traffic was booming. Lock and Dam Number 26 near my back door was finished by January 1938, and during its first year of operation 1,400,000 tons of shipping passed through that lock alone.

The new dams galvanized commercial traffic on the Upper Mississippi. In high-water periods of spring and early summer the gates of the big channel dams were simply raised to provide an open river; in low-water periods of summer the gates were closed or partially closed, impounding the River so that a main shipping channel at least nine feet deep could be maintained. For the first time there was dependable channel depth to accommodate heavy, deep-draft barges with their immense payloads.

The effects on the old, natural, free-flowing Upper Mississippi were even more spectacular. Its entire character was changed. When the Upper Mississippi River Fish and Wildlife Refuge was created in 1924, the River bottoms consisted of heavily wooded islands with hundreds of lakes and ponds scattered through them. Many of these dried up completely during summer, and fish-rescue work in the shrinking ponds was a prime conservation activity of the day. (And a miserable one, too, for the men laboring with seines in the heavy heat and glutinous mud of those landlocked ponds. The tombstone of one man who died after several years on a fish crew was engraved: "Died of overwork at the fish hatchery.")

Upper River marshes in the pre-dam days were largely limited to the shores of backwaters, deep sloughs, and the ponds of wooded islands, and even there the marsh flora was limited in variety and extent. As a result, most of the waterfowl found in the Upper Mississippi Refuge were diving ducks, especially scaup. Puddle ducks such as mallards were relatively limited simply because the marshy habitats they require were in such limited supply. In the spring, according to veteran biologist Dr. William Green, late of the United States Fish and Wildlife Service, a great variety of waterfowl would move up the River when the bottoms were flooded, but hunting was relatively limited and comparatively few hunters frequented the River as a whole. There would be shooting when a flight was on, but the action would soon drop off since there were no food supplies to hold the birds.

The first of the new impoundments on the Upper Mississippi Refuge was filled in 1935, and by the time the last was filled in 1939 an entirely new refuge had been created. No longer did the River fluctuate wildly between the floods of spring and the dry-outs of late summer; now the River was more or less stabilized, with much of it held in a more or less permanent condition of high water. The only places where the old condition of wooded islands and deep sloughs was now found was in the upper end of each impounded portion. Toward the center of the pool the effects of impoundment began to be apparent, with the River backing up over islands, old meadows, and low-lying bottomlands to form broad marshes that never existed before. The lower part of each pool, just above its dam, was deep open water that might lack marsh habitat for mallards and wading birds but would be the favorite haunt of such diving ducks as scaup, redheads and canvasback.

The best of the new habitats tended to be in the upper end of the refuge north of the Wisconsin River, especially in Pools 4 to 7. Since those pools are short they were more affected by impoundment than are pools farther downriver. Pools 10 to 15 average over thirty miles in length, and their upper ends are essentially in the old natural river condition. They tend to have more deep-water areas in which marsh conditions do not readily develop.

At the upper end of the refuge, where many of the islands and timbered ridges had been clear-cut before flooding, and the bottomlands heavily grazed, vast beds of Muhlenberg's smartweed appeared the year after the dam gates were closed. For several years it was a rich source of wildfowl food, and then it inexplicably vanished. But other rich aquatics took its place—at least seven other species of smartweeds, leafy pondweed, sago pondweed, wild rice, rice cutgrass, wild celery, and the rich beds of American pondweed which, thirty years ago, was probably the most important of all to waterfowl. An outstanding duck food, it does best in water not over thirty inches deep and may grow so thickly in some bays and backwaters that boating there is almost impossible.

In a 1954 report, Dr. Green noted that excellent marshes and aquatic habitats had developed where there were formerly wooded islands and dry marshes. No longer was there any lack of marsh and aquatic plants; fish-rescue work was a thing of the past, and the new river had become excellent habitat for furbearers and waterfowl.

Bill Green wasn't alone in his praise of the new Upper Mississippi. The late Dr. Ira Gabrielson, former chief of the United States Biological Survey, wrote in 1943 that no single conservation organization could have helped wildlife as much as the new nine-foot channel. It had a parallel effect on fish and fishing as well.

Don Edlen of the Sabula (Iowa) Fisheries Station reported sweeping changes—most for the better. Backwaters, he noted, were greatly expanded and stabilized, and now provided dependable, greatly expanded spawning habitats. Public access to these backwaters was good and would become even better, adding a whole new dimension to the River's sportfishery, as would the deep, wild, highly aerated water below the channel dams that would draw spring and fall concentrations of walleyes—and sportsmen.

For several seasons, commercial fishermen were somewhat puzzled and frustrated by the River's new form. Fish had been widely scattered throughout the greatly expanded water areas, with the result that populations of fish per acre were unusually low. But within three years the new spawning situations were beginning to tell; nature abhors any vacuum, and many river fish were now occupying the vast new habitats. The professional fishermen soon began figuring out those new habitats and how fish used them— and in 1957 Edlen told me that commercial fishermen were making three times as much money as they had before the advent of the nine-foot channel.

Still and all, it was to prove a mixed bag.

Bill Green recently noted that for thirty years "conditions remained excellent" on the Upper Mississippi River Wildlife and Fish Refuge, but "once the pools became permanently established the normal deterioration associated with stabilized water areas gradually began." In other words, broad areas of the new chain of river lakes began choking with silt and sand pouring down through tributary streams from intensively farmed uplands.

A stream's capacity to hold nonbuoyant particles in suspension is directly proportional to its velocity. As the velocity decreases, so does the stream's capacity to suspend and move silt and sand, and those materials will normally settle out. The original Upper Mississippi, with the appreciable gradient usually associated with upper reaches of rivers, had a pretty good current. There were times when its midstream velocity might be five miles per hour or even more when the River was in spate, and it did a pretty fair job of moving burdens of silt and sand on downstream. With the advent of the channel dams, the average current velocity was greatly reduced—and so was the River's ability to transport particulate matter. This is most apparent in the midreaches of each pool, where the "natural" river begins to merge into the more slow-moving impounded section. It is there that the silt and sand borne in the current begin to settle out as the River slows, and the same factors that helped create some of the new wetland habitats and their rich beds of aquatic vegetation now began to undo the gains.

Predictably, the big channel dams have formed catch basins for the silt that pours down from the uplands. Such erosion is a natural process that cannot be checked, but it has run far beyond natural bounds. The natural Upper Mississippi was self-renewing to a degree, and the same natural processes that tended to fill some backwaters and side channels also created new ones in an ecosystem that supported a wide diversity of fish and wildlife. Long reaches of river were regularly scoured and flushed, their substrates of rock and gravel swept free of silt and debris to form rich habitats for bottom fauna. With the slowing of the River by the new dams, sedimentation was greatly advanced. The backwaters and side channels have been filled at average rates from one-half an inch to two inches per year. At that rate, one report concluded, "within a century most of the Upper Mississippi River system will consist of a main channel bordered by dry land or shallow marsh."

The source of the problem, of course, is the general, "non-point" pollution from farmlands along the River's watershed. A report of the Great River Environmental Action Team (GREAT) noted that "the most pervasive and damaging problem for the Upper Mississippi River is soil eroded from the basin's uplands which settles the river's backwaters."

Twenty years ago we were told that every bushel of corn produced in the Upper Midwest costs a bushel of eroded topsoil. In some Iowa counties along the River the farmlands bleed an average of over thirteen tons of topsoil into the Mississippi annually—almost three times acceptable "normal" loss. Not all farmlands on the watershed are eroding seriously, of course, and there are many careful, conscientious farmers who jealously husband their topsoil. In my personal experience, though, the typical midwestern farmer still practices the sort of predatory farming that has always characterized prairie agriculture. More often than not, he is a worried man who has overextended himself on the advice of marketing specialists and policymakers. In a vain effort to keep from drowning in red ink, he has farmed more and more marginal land, clearing wooded hillsides, plowing slopes, and greatly increasing the scale of his operations. The feverish need to balance his books has driven him to commit maximum acreage to intensive use, which demands newer and heavier equipment and more expensive pesticides and fertilizers, driving him even deeper into debt. His compulsion to increase crop production has also greatly increased impacts on the resource base itself—and today that loss to erosion may be *two* bushels of topsoil to each bushel of corn produced. The lower portions of some of our favorite smallmouth-bass streams, those bright little waters that exemplified so much quality and joy—the Zumbro, Root, Upper Iowa, Cannon, Black and Wapsipinicon— now run turbid and thick with the silts being wasted in behalf of needless grain production. And the rich beds of wild celery in Pool 8, an aquatic plant so cherished by the lordly canvasback duck that they even share the same scientific name, have undergone a catastrophic decline for several reasons—the most serious, by far, being siltation.

Many of the rich backwaters that resulted from the new channel dams have now all but vanished, and deep sloughs that were once prime fishing spots can no longer be traveled by boat at all. "Point pollution," caused by discharges of municipal sewage and industrial wastes, can often be identified and attended to, but the "non-point pollution" of agricultural chemicals and soil erosion is a terribly hard thing to address. The first type has been somewhat reduced in recent years, but siltation is worse than ever. Much depends, of course, on whom you listen to. In the 1982 *Comprehensive Master Plan for the Management of the Upper Mississippi River Systems,* state and federal fish and game agencies contributing to the study noted that in the past thirty-five years a fourth of the open-water area of the Upper River's impoundments had been converted to marsh or terrestrial habitat, and that lakes formerly up to fifteen feet deep were now less than two feet deep. They predicted that at current sedimentation rates "substantial backwater areas will be eliminated in 50 years or less." It was also pointed out that the channel dams had "essentially stopped the meandering processes of the river which create backwaters and side channels. By changing flow in certain regimes and diverting water to the main channel, these (the channel dams and wing dams) contributed to accelerated sedimentation in backwaters and side channels and limited new formation."

While this master plan was still in draft, a public hearing was held in Saint Louis late in 1981 to provide input and to test the political climate. As is usual with such hearings, there was more heat than light.

One of the livelier presentations was that of Guy Jester, president of the Association for the Improvement of the Mississippi River. Jester was formerly district engineer, Saint Louis District, United States Army Corps of Engineers. Serving from 1971 to 1973—the time that saw Earth Day and the dawn of the Environmental seventies—it is reasonable to assume that Colonel Jester had a bellyful of hot-eyed environmentalists early on. At the 1981 hearing it was apparent that he still hadn't digested them.

> The environmental studies that I have been able to see and in this report are clearly biased against transportation, they are replete with unsupported statements and opinions. The write-up . . . beginning at results of the environmental to the end of the paragraph procedural value should be deleted as untrue and pure speculation. The value of the overall report should not be destroyed by such tripe.
>
> There are such statements as "prior to navigation development the Mississippi River was itself renewing." Hogwash. This is blatently [sic] untrue. . . . Flora and fauna was [sic] suffering and dying, the stream was becoming shallower and shallower, new back channels and side channels were not being created as this statement says, they were being filled rapidly. The navigation improvements actually prolonged the life of this river if you will look at the history of what has gone on.

But although the navigation improvements may have expanded certain wetland habitats along the River, it was anything but a permanent blessing.

Few men know the living River any better than Dr. Richard

Sparks of the Illinois Natural History Survey. A resourceful, deeply dedicated field biologist whose research on the effects of barge traffic has sometimes been short-circuited by commercial interests, he takes issue with the corps's contention that present-day losses of those expanded habitats is nothing worse than "just going back to the amounts of habitat we once had." He recently wrote:

> Man no longer tolerates a free-flowing river that, when flooded, would carve new channels and leave behind new backwaters that would become productive habitat for waterfowl, furbearers and other creatures that thrive in a wetland and bottomland environment.
>
> Now we've put our farms and industries in certain areas along the river, and we've established a navigation channel in others, and we just can't tolerate a river that keeps moving around. We're not allowing the river to create any new areas, new habitat, as it once did. So for awhile we may still be ahead in total habitat acreage. But for how long?

In considering the ills that plague the Upper River it is hard to assign the greatest degree of guilt. Many corps watchers believe that the Corps of Engineers is culpable for the steady degradation of the Upper Mississippi; others draw a bead on the towboat industry or agribusiness. But there is enough blame to go around, and all can be tarred with the same brush. The corps undeniably started it all with the channel dams. In their own defense, the colonels usually protest that they are simply servants of the people, responding to mandates of the Congress. However, the corps is regarded by many as the most powerful special-interest lobby in Washington—proprietor of a virtually bottomless pork barrel and the darling of senators and congressmen seeking dams, reservoirs, and channelized streams for certain folks back home. But while I can get as exercised about the corps as anyone, I know there are always men and women in the corps who tend to feel as I do, and do what they can to temper projects and policies with sound resource management.

Modern barge lines can be held responsible for at least part of the problems confronting the Upper River. Sired by the Corps of Engineers with the River's canalization, they are still dependent on their parent and could probably not exist if not subsidized with public money. Token user taxes are levied on the barge lines, but such revenue is barely a scratch on the huge sums needed to maintain these inland waterways. There is no denying that the trucking, railway, and airlines industries have also been in the federal trough, but none to the degree of the barge lines. The nine-foot navigation channel of the Upper Mississippi River was simply created, and is maintained, for the sole benefit of commercial river traffic. The various industries benefiting from this largess claim that such public waterways are of enormous economic importance to the nation and well worth the cost of support, although it is questionable whether the public at large ever gets its money's worth—especially in view of the problems that develop.

One of these is the incessant maintenance of the navigation channel by dredging, and disposal of the dredge spoil. As much as 2,261,000 cubic yards of material has been dredged in one year from the head of navigation through Pool 10. Getting rid of this stuff is a never-ending problem, for the most convenient disposal sites may be certain backwaters that are of great value to fish and wildlife. And even if the disposal sites have little habitat value, the process of dredging can increase water turbidity, destroy mussel beds (in 1976 at Prairie du Chien a corps dredge sucked up a hundred thousand cubic yards of sand that held over 1,700 mussels of 33 kinds, including 60 of the endangered Higgins eye), reduce dissolved oxygen, and release toxic wastes that had been locked into polluted sediments.

Then there is the direct effect of the big towboats whose powerful engines churn bottom sediments and create turbidity that has scarcely a chance to clear before another towboat renews it. Bank erosion is accelerated, and the rate of filling in backwaters is increased.

One of the hottest local issues where I live is the problem of barge fleeting. That is, where and how to hold the big strings of barges as they await their turn to lock through a channel dam. Lock and Dam 26 at Alton is a particular bottleneck to barge traffic because it lies just below the junction of the Illinois and Mississippi rivers and serves both major waterways. Number 26 catches all the barge traffic bound upstream and down from both rivers and is simply unable to serve it efficiently. There are long delays in which strings of barges are stalled above and below such dams, angering towboat captains and riverside residents alike.

Our friends Bernell and Forrest Narup have a fine home beside the River at Hastings Landing on the west side of Calhoun County, Illinois, about three miles below Lock and Dam 25. On their riverbank are several large cottonwoods that are ideal for the tying off of barges while waiting for traffic jams to clear. The huge hawsers required for this can make short work of even the biggest cottonwood by girdling it, or may snap off a tree or simply uproot it. The Narups' cottonwoods are private property and so posted, but that may make little difference to a towboat captain in need of a stopping place.

One afternoon a couple of years ago Forrest returned from a duck hunt to find a string of barges tied to one of his huge trees. He went down to the riverbank where he yelled for a while and then, failing to get anyone's attention, returned to his tool shed for an axe.

"I started cutting away at that big hawser and *then* I got some attention!" he told me. "Either the captain or the mate, I don't know which, came out and ordered me to quit cutting his line. It was a new line that had cost several hundred dollars, he said, and he wasn't going to have it ruined. When I didn't stop he reached into his belt and pulled out the biggest .45 I'd ever seen. Said he was going to blow my head off. I was just as mad as he was, maybe even madder, and said that if he didn't put that gun away he was going to get his head split with an axe.

"If you've never tried to cut through a four-inch poly hawser with a dull axe, you just can't appreciate the situation," Forrest said. "Especially with an armed man threatening to shoot you. Cutting through that big line seemed to take forever, but it finally parted—and so did I."

"Did you report this to the authorities?"

"I did. The sheriff was willing to ticket the barge lines for trespass, but the state's attorney wasn't willing to prosecute. So I wrote the Coast Guard and the Corps of Engineers and my congressman, and then they all began exchanging letters, and after several years we've built up quite a file."

"Has it done any good?"

"Not that you'd notice."

"What if it happens again?"

"Well," said Forrest, "next time the axe will be sharper!"

Not far downriver, just above the Alton dam, the problem of barge fleeting is at its worst. I have seen towboats and their strings of barges lined up along the riverbank almost jackstaff to fantail for ten miles. In recent years there has been considerable pressure to designate parts of this stretch as fleeting sites for the assembling and shuffling of barge groups, with permanent anchors and tie-off facilities that would transform some of the now-empty riverfront into an industrial area.

This is an especially sensitive issue because such fleeting areas would flank the Great River Road that runs along the bank from Alton to Grafton, seventeen miles of Mississippi River panorama at its best—a broad new four-lane highway with the mile-wide Mississippi on one side and high limestone palisades on the other. No other stretch of the Great River Road, from Itasca to New Orleans, offers such ready access to such a superb riverscape to so many people. Small wonder that many of our local folks are wroth at the thought of industrializing this stretch of riverfront. Others are more complacent about it. Not long ago one of my friends in Grafton said: "Maybe barge fleeting won't exactly beautify the River, but we've got to make *some* concessions to progress." To which I could only add that for forty years I've been making concessions to progress. I have been neatly conceded out of many of my best fishing, hunting, and boondocking places and have grown chronically suspicious of anyone thumping the tub for what usually turns out to be progress of the unilateral kind.

The solution to the barge fleeting and traffic jam problem at Alton is at hand, but could preface even greater headaches. With the present volume of river traffic the two old locks at Dam 26 are clearly inadequate. And since those locks and the dam itself are said to be in wretched condition, they are being replaced a couple of miles downstream by a new Lock and Dam 26 that is expected to end up costing over $800 million by the scheduled 1989 completion date. It has occurred to some of us that for that kind of money freight could be airmailed from Saint Louis to Saint Paul for the next thirty years or so.

The new lock and dam beggar most superlatives. It is the biggest project ever undertaken by the Corps of Engineers, which is a pretty good superlative in itself. First, a protective cofferdam was needed—a steel enclosure built out into the River and then pumped dry to provide a place for the actual construction to begin. This meant a double wall of interlocking steel sheet piles enclosing twenty-five acres of riverbed. Over nineteen thousand of the ninety-five-foot steel pilings were driven to near bedrock, down through recent sands, through gravels of the Wisconsin glaciation, and into rubble left by glaciers of the Illinois Ice Age up to 125,000 years ago. The spaces between the walls of piling were then filled with over two million tons of sand and rock to form a roadbed sixty feet wide. The cofferdam completed, the enclosed area was pumped dry by the ten powerful turbine pumps that expel over 100 million gallons of seepage each twenty-four hours—the equivalent of a 1¼-inch stream of water reaching from here to the moon. With all this accomplished, and twenty-five acres of riverbed exposed, work on the spillway section of the actual dam began.

For years I've heard wild tales about the huge fish that were stranded when the cofferdam of the existing Lock and Dam 26 was pumped dry in the late 1930s, and wondered if the same would occur at the new site. However, I was even more interested in seeing twenty-five acres of mainstem Mississippi River bottom revealed.

"You'll be disappointed," said Jim Bissell, the project's resident engineer. "It'll be pretty well torn up by equipment and certainly won't be the kind of undamaged riverbed you're interested in seeing."

"Well, how about any fish rescue efforts?" I asked, hoping for second best. "There could be lots of fish trapped in there."

"We haven't been contacted by any Fish and Game people," Bissell answered. "As far as I know, there are no plans for any fish rescue work. There might be a lot of fish trapped inside that cofferdam, but some of the men figure that when it's pumped dry we'll find just one B-I-G catfish!" I like Jim Bissell. He didn't end up with a monster catfish after all, but he's going to end up with one devil of a big channel dam.

Instead of trucking in concrete, the corps simply built its own concrete factory on the site. Within the protecting walls of the cofferdam that rise fifty feet above the riverbed a small army of construction workers is raising the massive concrete dam piers to which the six gates will be fixed. With this major spillway section completed, work on phase two—the construction of the main lock section—will begin. It is this feature, twelve hundred feet long and one hundred ten feet wide, that has ignited much of the controversy over the huge project.

Two miles upstream, the six-hundred-foot main lock chamber of old Dam 26 has a maximum capacity of nine barges and their towboat. Tows with more barges than that (and many of the strings now contain over nine barges) must be laboriously broken apart and locked through in sections, a process that can take up to two hours—hence the traffic jams. The new structure will permit up to fifteen barges to pass through in one lockage that may require less than twenty minutes. In addition, an auxiliary lock chamber of six hundred feet has been proposed. When the new Lock and Dam 26 enter service, the old one upstream will be demolished. The most serious commercial bottleneck on the Upper Mississippi will be removed—and a Pandora's chest of environmental problems may be opened.

What will happen when the new lock creates a capacity bulge that results in upstream traffic jams, as it is sure to do? Will the corps proceed to enlarge the locks in other dams, as it did on the Ohio River? There is little doubt that the new Lock and Dam 26 will increase barge-line use of a river that is already felt to carry far too much traffic for its own good. But the central fear of environmentalists stems from the fact that the new lock will have eighteen-foot sills that can accommodate deeper, heavier barges and ultimately lead to the creation of a twelve-foot navigation channel in the Upper Mississippi.

Corpsmen deny this, pointing out that the eighteen-foot sill is needed in winter for barges that have accumulated thick buildups of ice on their bottoms. An eighteen-foot sill depth, they say, is quite proper for the winter use of barges with a nine-foot draft. In an advertising supplement of the *St. Louis Post-Dispatch*, sponsored by a group of barge-line companies, major shippers, and other interests, the corps was quoted as saying "for certain that there is no 12-foot channel planned for the Upper Mississippi River; only Congress, through its authorizing and appropriating powers, has the final authority to instigate and approve this matter." Several years earlier, however, a spokesman of the Saint Louis Corps District told the Society of Civil Engineers that an eighteen-foot lock sill "will provide sufficient depth for tows with a twelve-foot draft plus additional clearance for normal operation and deposits of ice, frozen soil and debris which frequently cling to the bottoms of barges during winter operations."

Whether or not the colonels are just biding their time to create a twelve-foot navigation channel is a matter of conjecture. But there is little doubt that such a project could wreak havoc with the wild values that are in growing jeopardy along the Upper Mississippi. There are only two basic ways to deepen the channel: raise the River an average of at least three feet above its present level, or dredge the existing channel deeper.

The first method would involve the building of new levee systems and raising old ones. Much of the Upper River would begin to approximate a channeled ditch between embankments—the condition that prevails on most of the Lower Mississippi. In addition to the immediate damage this could visit on backwater habitats so valuable to fish and wildlife, many undeveloped bottomlands would also be lost. Low country now in flood timber and wetlands could be safely converted to soybean fields and even industry if it lay secure behind levees designed to hold back a hundred-year flood.

As it is, the River's bed has already been raised several feet with the sediments that have accumulated since the coming of the channel dams. It isn't just coincidence that some of our most spectacular flood crests have occurred since then; the River may not be carrying all that much more floodwater but is simply carrying it in a streambed with less capacity.

Deepening the channel could also ring the tocsin for valuable backwaters. It has been estimated that in order to create a navigation channel twelve feet deep and four hundred feet wide, with additional width in many bends, 20 percent of the River between

Grafton and Minneapolis would need dredging. This has been translated to mean about twenty-three-million cubic yards of spoil removed from the channel—and disposition of that spoil somewhere in the shallows, backwaters, or banks of the Upper Mississippi. Either way, to say nothing of the damaging effects of a fourfold increase in barge traffic, the River loses.

Afflicted as they are with chronic paranoia, environmentalists must surely bore the socks off the general public. Forever crying havoc and warning of dire natural calamity, embracing pupfish and furbish louseworts and obscure causes that are of profound disinterest to most people, their zeal often sets them apart as odd people that should not be taken too seriously. Except, of course, when their queer ways tend to Impede Progress.

It is not hard to appreciate that side of it. Few people are more tedious than a solemn, zealous environmentalist who's aflame with righteousness. But there is another side—one occupied by thoughtful, experienced men and women who have paid their dues with years of outdoor mileage, long training, and often both. And when someone like my old friend Jim Mayhew, fisheries chief of the Iowa Conservation Commission, or a commercial fisherman like Worth Emmanuel of Maiden Rock, Wisconsin, opines that the Upper Mississippi River is dying, I take him very seriously indeed.

Aside from the fact that people with a stake in quality riverscapes are confused and somewhat wearied by the incessant pulling and hauling and the growing complexities of the situation, there is the question of relative price tags. The commercial users of the River—the barge lines, shippers, manufacturers, port authorities, and most especially the Corps of Engineers—have grown adept at producing numbers to support their exploitation of the Upper Mississippi. And although professional resource managers can come up with approximate values for catchable fish and huntable wildlife, they are hard put to present acceptable values for aesthetics. It just isn't enough to say that a clean-flowing, life-sustaining, joy-giving Upper Mississippi is, like virtue, sufficient reward unto itself, for it is more likely to be an object of exploitation in a society where both virtue and living rivers are held in amused contempt by broad sectors of politics and business.

In theory, resource conservation agencies have a voice in how the Upper Mississippi will be used. In practice, such agencies are subordinate to administrations of state and federal governments—and those administrations are highly sensitive to the demands of shippers who want as much heavy towboat traffic as possible, and an Army Corps of Engineers that is dedicated to clearing the way for such traffic. And if these do not actually hold title to the Mississippi, they are surely the proprietors.

In her book *Rivers That Meet in Egypt,* Barbara Burr Hubbs tells of a lanky southern Illinois farmer riding a mule down to a Mississippi River landing where a steamboat was docked. He asked to see the captain. No one else would do. It was necessary to speak to the Great One himself. When that kingly personage finally deigned to appear on the texas deck, the farmer asked: "Jest wanted to know, kin my mule drink out of your river?"

A lot of us have been wondering the same thing. Our mules are getting almighty thirsty.

Far downriver, at the USAE Waterways Experiment Station at Vicksburg, the Army Corps of Engineers operates a scale model of the Mississippi. Carefully detailed in concrete, it simulates the hydrology of the River and occupies several acres.

Some years ago my wife fashioned her own model of the Mississippi—one that can be placed on a tabletop but is no less valid than that designed by the engineers. It is the figure of a burly old Indian, head shaven but for a scalp lock. He is stern, frowning slightly, his powerful body straining forward in great and patient effort. Clinging to his broad back are several small figures—a frog, a water nymph, a fish.

"Tell me about it," I asked.

"It is the Old River and his children," Dycie said. "He is taking them away."

"Where are they going?"

"No one can say," she replied. "Perhaps back into the past where they began. Or maybe into the far future. But it can only be away from here and now, into a time that cares."

For a complete list of books available from Penguin in the United States, write to Dept. DG, Penguin Books, 299 Murray Hill Parkway, East Rutherford, New Jersey 07073.

For a complete list of books available from Penguin in Canada, write to Penguin Books Canada Limited, 2801 John Street, Markham, Ontario L3R 1B4.